BROOKLANDS BOOKS

BMW

M SERIES

PERFORMANCE PORTFOLIO

1976·1993

Compiled by
R. M. CLARKE

ISBN 185520 2107

Brooklands Books Ltd.
PO Box 146, Cobham, KT11 1LG
Surrey, England.

Printed in Hong Kong

BROOKLANDS
BOOKS

BROOKLANDS ROAD TEST SERIES

Abarth Gold Portfolio 1950-1971
AC Ace & Aceca 1953-1983
Alfa Romeo Alfasud 1972-1984
Alfa Romeo Alfetta Coupés GT. GTV. GTV6 1974-1987
Alfa Romeo Giulia Berlinas 1962-1976
Alfa Romeo Giulia Coupés Gold Portfolio 1963-1976
Alfa Romeo Giulia Coupés 1963-1976
Alfa Romeo Giulietta Gold Portfolio 1954-1965
Alfa Romeo Spider Gold Portfolio 1966-1991
Alfa Romeo Spider 1966-1991
Allard Gold Portfolio 1937-1959
Alvis Gold Portfolio 1919-1967
American Motors Muscle Cars 1966-1970
Armstrong Siddeley Gold Portfolio 1945-1960
Aston Martin Gold Portfolio 1972-1985
Austin Seven 1922-1982
Austin A30 & A35 1951-1962
Austin Healey 100 & 100/6 Gold Portfolio 1952-1959
Austin Healey 3000 Gold Portfolio 1959-1967
Austin Healey Sprite 1958-1971
BMW Six Cyl. Coupés 1969-1975
BMW 1600 Collection No.1 1966-1981
BMW 2002 1968-1976
BMW 316, 318, 320 (4 cyl.) Gold Portfolio 1975-1990
BMW 320, 323, 325 (6 cyl.) Gold Portfolio 1977-1990
BMW 5 Series Gold Portfolio 1981-1987
Bristol Cars Gold Portfolio 1946-1992
BMW M Series 1976-1993
Buick Automobiles 1947-1960
Buick Muscle Cars 1965-1970
Cadillac Automobiles 1949-1959
Cadillac Automobiles 1960-1969
Chevrolet Camaro Z28 & SS 1966-1973
Chevrolet Camaro & Z28 1973-1981
High Performance Camaros 1982-1988
Camaro Muscle Portfolio 1967-1973
Chevrolet 1955-1957
Chevrolet Corvair 1959-1969
Chevrolet Impala & SS 1958-1971
Chevrolet Muscle Cars 1966-1971
Chevelle & SS Muscle Portfolio 1964-1972
Chevy Blazer 1969-1981
Chevy El Camino & SS 1959-1987
Chevy II Nova & SS 1962-1973
Chevrolet Corvette Gold Portfolio 1953-1962
Chevrolet Corvette Sting Ray Gold Portfolio 1963-1967
Chevrolet Corvette Gold Portfolio 1968-1977
High Performance Corvettes 1983-1989
Chrysler 300 Gold Portfolio 1955-1970
Chrysler Valiant 1960-1962
Citroen Traction Avant Gold Portfolio 1934-1957
Citroen 2CV 1948-1988
Citroen DS & ID 1955-1975
Citroen SM 1970-1975
Cobras & Replica 1962-1983
Shelby Cobra Gold Portfolio 1962-1969
Cobras & Cobra Replicas Gold Portfolio 1962-1989
Cunningham Automobiles 1951-1955
Daimler SP250 Sports & V-8 250 Saloon Gold Portfolio 1959-1969
Datsun Roadsters 1962-1971
Datsun 240Z 1970-1973
Datsun 280Z & ZX 1975-1983
De Tomaso Collection No. 1 1962-1981
Dodge Charger 1966-1974
Dodge Muscle Cars 1967-1970
Dodge Viper on the Road
Excalibur Collection No. 1 1952-1981
Facel Vega 1954-1964
Ferrari Cars 1946-1956
Ferrari Dino 1965-1974
Ferrari Dino 308 1974-1979
Ferrari 308 & Mondial 1980-1984
Ferrari Collection No. 1 1960-1970
Motor & T&CC Ferrari 1966-1976
Motor & T&CC Ferrari 1976-1984
Fiat-Bertone X1/9 1973-1988
Fiat Pininfarina 124 & 2000 Spider 1968-1985
Ford Consul, Zephyr, Zodiac Mk.I & II 1950-1962
Ford Zephyr Zodiac Executive Mk.III & Mk.IV 1962-1971
Ford Cortina 1600E & GT 1967-1970
High Performance Capris Gold Portfolio 1969-1987
Capri Muscle Portfolio 1974-1987
High Performance Fiestas 1979-1991
High Performance Escorts Mk.I 1968-1974
High Performance Escorts Mk.II 1975-1980
High Performance Escorts 1980-1985
High Performance Escorts 1985-1990
High Performance Sierras & Merkurs Gold Portfolio 1983-1990
Ford Automobiles 1949-1959
Ford Fairlane 1955-1970
Ford Ranchero 1957-1959
Thunderbird 1955-1957
Thunderbird 1958-1963
Thunderbird 1964-1976
Ford Falcon 1960-1970
Ford GT40 Gold Portfolio 1964-1987
Ford Bronco 1966-1977
Ford Bronco 1978-1988
Holden 1948-1962
Honda CRX 1983-1987
Hudson & Railton 1936-1940
Jaguar and SS Gold Portfolio 1931-1951
Jaguar XK120, XK140, XK150 Gold Portfolio 1948-1960
Jaguar Mk.VII, VIII, IX, X, 420 Gold Portfolio 1950-1970
Jaguar 1957-1961
Jaguar Mk.2 1959-1969
Jaguar Cars 1961-1964
Jaguar E-Type Gold Portfolio 1961-1971
Jaguar E-Type 1966-1971
Jaguar E-Type V-12 1971-1975
Jaguar XJ12, XJ5.3, V12 Gold Portfolio 1972-1990
Jaguar XJ6 Series II 1973-1979
Jaguar XJ6 Series III 1979-1986
Jaguar XJS Gold Portfolio 1975-1990
Jeep CJ5 & CJ6 1960-1976
Jeep CJ5 & CJ7 1976-1986
Jensen Cars 1946-1967
Jensen Cars 1967-1979
Jensen Interceptor Gold Portfolio 1966-1986
Jensen Healey 1972-1976
Lagonda Gold Portfolio 1919-1964
Lamborghini Cars 1964-1970
Lamborghini Countach & Urraco 1974-1980
Lamborghini Countach & Jalpa 1980-1985
Lancia Fulvia Gold Portfolio 1963-1976
Lancia Stratos 1972-1985
Land Rover Series I 1948-1958
Land Rover Series II & IIa 1958-1971
Land Rover Series III 1971-1985
Land Rover 90 & 110 1983-1989
Lincoln Gold Portfolio 1949-1960
Lincoln Continental 1961-1969
Lincoln Continental 1969-1976
Lotus & Caterham Seven Gold Portfolio 1957-1989
Lotus Elite 1957-1964
Lotus Elite & Eclat 1974-1982
Lotus Elan Gold Portfolio 1962-1974
Lotus Elan Collection No. 2 1963-1972
Lotus Cortina Gold Portfolio 1963-1970
Lotus Europa Gold Portfolio 1966-1975
Lotus Turbo Esprit 1980-1986
Motor & T&CC on Lotus 1979-1983
Marcos Cars 1960-1988
Maserati 1965-1970
Maserati 1970-1975
Mazda RX-7 Collection No. 1 1978-1981
Mercedes Benz Cars 1949-1954
Mercedes Benz Competition Cars 1950-1957
Mercedes Benz Cars 1954-1957
Mercedes Benz Cars 1957-1961
Mercedes 190 & 300 SL 1954-1963
Mercedes 230/250/280SL 1963-1971
Mercedes Benz SLs & SLCs Gold Portfolio 1971-1989
Mercedes S & 600 1965-1972
Mercedes S Class 1972-1979
Mercury Muscle Cars 1966-1971
Metropolitan 1954-1962
MG Gold Portfolio 1929-1939
MG TC 1945-1949
MG TD 1949-1953
MG TF 1953-1955
MG Cars 1959-1962
MGA & Twin Cam Gold Portfolio 1955-1962
MG Midget 1961-1980
MGB Roadsters 1962-1980
MGB MGC & V8 Gold Portfolio 1962-1980
MGB GT 1965-1980
Mini Cooper Gold Portfolio 1961-1971
Mini Muscle Cars 1961-1979
Mini Moke 1964-1989
Mopar Muscle Cars 1964-1967
Morgan Three-Wheeler Gold Portfolio 1910-1952
Morgan Plus 4 & Four 4 Gold Portfolio 1936-1967
Morgan Cars 1960-1970
Morgan Cars Gold Portfolio 1968-1989
Morris Minor Collection No. 1 1948-1980
Shelby Mustang Muscle Portfolio 1965-1970
Mustang Muscle Cars 1967-1971
High Performance Mustang IIs 1974-1978
High Performance Mustangs 1982-1988
Oldsmobile Automobiles 1955-1963
Oldsmobile Cutlass & 4-4-2 1964-1972
Oldsmobile Muscle Cars 1964-1971
Oldsmobile Toronado 1966-1978
Opel GT 1968-1973
Packard Gold Portfolio 1946-1958
Pantera Gold Portfolio 1970-1989
Panther Gold Portfolio 1972-1990
Plymouth Barracuda 1964-1974
Plymouth Muscle Cars 1966-1971
Pontiac Tempest & GTO 1961-1965
Pontiac Muscle Cars 1966-1972
Pontiac Firebird & Trans-Am 1973-1981
High Performance Firebirds 1982-1988
Pontiac Fiero 1984-1988
Porsche 356 1952-1965
Porsche Cars in the 60's
Porsche Cars 1960-1964
Porsche Cars 1964-1968
Porsche Cars 1968-1972
Porsche Cars 1972-1975
Porsche 911 1965-1969
Porsche 911 1970-1972
Porsche 911 1973-1977
Porsche 911 Carrera 1973-1977
Porsche 911 Turbo 1975-1984
Porsche 911 SC 1978-1983
Porsche 914 Collection No. 1 1969-1983
Porsche 914 Gold Portfolio 1969-1976
Porsche 924 Gold Portfolio 1975-1988
Porsche 928 1977-1989
Porsche 944 1981-1985
Range Rover Gold Portfolio 1970-1992
Reliant Scimitar 1964-1986
Riley 1.5 & 2.5 Litre Gold Portfolio 1945-1955
Rolls Royce Silver Cloud 1955-1965
Rolls Royce Silver Cloud & Bentley 'S' Series Gold Portfolio 1955-1965
Rolls Royce Silver Shadow 1965-1981
Rover P4 1949-1959
Rover P4 1955-1964
Rover 3 & 3.5 Litre Gold Portfolio 1958-1973
Rover 2000 & 2200 1963-1977
Rover 3500 1968-1977
Rover 3500 & Vitesse 1976-1986
Saab Sonett Collection No.1 1966-1974
Saab Turbo 1976-1983
Studebaker Gold Portfolio 1947-1966
Studebaker Hawks & Larks 1956-63
Avanti 1962-1990
Sunbeam Tiger & Alpine Gold Portfolio 1959-1967
Toyota MR2 1984-1988
Toyota Land Cruiser 1956-1984
Triumph TR2 & TR3 1952-1960
Triumph TR4, TR5, TR250 1961-1968
Triumph TR6 Gold Portfolio 1969-1976
Triumph TR7 & TR8 1975-1982
Triumph Herald 1959-1971
Triumph Vitesse 1962-1971
Triumph Spitfire Gold Portfolio 1962-1980
Triumph 2000, 2.5, 2500 1963-1977
Triumph GT6 1966-1974
Triumph Stag 1970-1980
TVR Gold Portfolio 1959-1990
VW Beetle Gold Portfolio 1935-1967
VW Beetle Gold Portfolio 1968-1991
VW Kubelwagen 1940-1975
VW Karmann Ghia 1955-1982
VW Bus, Camper, Van 1954-1967
VW Bus, Camper, Van 1968-1979
VW Bus, Camper, Van 1979-1989
VW Beetle Collection No.1 1970-1982
VW Scirocco 1974-1981
VW Golf GTI 1976-1986
Volvo PV444 & PV544 1945-1965
Volvo Amazon-120 Gold Portfolio 1956-1970
Volvo 1800 Gold Portfolio 1960-1973

BROOKLANDS ROAD & TRACK SERIES

Road & Track on Alfa Romeo 1949-1963
Road & Track on Alfa Romeo 1964-1970
Road & Track on Alfa Romeo 1971-1976
Road & Track on Alfa Romeo 1977-1989
Road & Track on Aston Martin 1962-1990
Road & Track on Auburn Cord and Duesenburg 1952-1984
Road & Track on Audi & Auto Union 1952-1980
Road & Track on Audi & Auto Union 1980-1986
Road & Track on Austin Healey 1953-1970
Road & Track on BMW Cars 1966-1974
Road & Track on BMW Cars 1975-1978
Road & Track on BMW Cars 1979-1983
Road & Track on Cobra, Shelby & Ford GT40 1962-1992
Road & Track on Corvette 1953-1967
Road & Track on Corvette 1968-1982
Road & Track on Corvette 1982-1986
Road & Track on Corvette 1986-1990
Road & Track on Datsun Z 1970-1983
Road & Track on Ferrari 1975-1981
Road & Track on Ferrari 1981-1984
Road & Track on Ferrari 1984-1988
Road & Track on Fiat Sports Cars 1968-1987
Road & Track on Jaguar 1950-1960
Road & Track on Jaguar 1961-1968
Road & Track on Jaguar 1968-1974
Road & Track on Jaguar 1974-1982
Road & Track on Jaguar 1983-1989
Road & Track on Lamborghini 1964-1985
Road & Track on Lotus 1972-1981
Road & Track on Maserati 1952-1974
Road & Track on Maserati 1975-1983
Road & Track on Mazda RX7 1978-1986
Road & Track on Mazda RX7 & MX5 Miata 1986-1991
Road & Track on Mercedes 1952-1962
Road & Track on Mercedes 1963-1970
Road & Track on Mercedes 1971-1979
Road & Track on Mercedes 1980-1987
Road & Track on MG Sports Cars 1949-1961
Road & Track on MG Sports Cars 1962-1980
Road & Track on Mustang 1964-1977
Road & Track on Nissan 300-ZX & Turbo 1984-1989
Road & Track on Peugeot 1955-1986
Road & Track on Pontiac 1960-1983
Road & Track on Porsche 1951-1967
Road & Track on Porsche 1968-1971
Road & Track on Porsche 1972-1975
Road & Track on Porsche 1975-1978
Road & Track on Porsche 1979-1982
Road & Track on Porsche 1982-1985
Road & Track on Porsche 1985-1988
Road & Track on Rolls Royce & Bentley 1950-1965
Road & Track on Rolls Royce & Bentley 1966-1984
Road & Track on Saab 1972-1992
Road & Track on Toyota Sports & GT Cars 1966-1984
Road & Track on Triumph Sports Cars 1953-1967
Road & Track on Triumph Sports Cars 1967-1974
Road & Track on Triumph Sports Cars 1974-1982
Road & Track on Volkswagen 1951-1968
Road & Track on Volkswagen 1968-1978
Road & Track on Volkswagen 1978-1985
Road & Track on Volvo 1957-1974
Road & Track on Volvo 1975-1985
Road & Track - Henry Manney at Large and Abroad

BROOKLANDS CAR AND DRIVER SERIES

Car and Driver BMW 1955-1977
Car and Driver BMW 1977-1985
Car and Driver on Cobra, Shelby & Ford GT40 1963-1984
Car and Driver on Corvette 1956-1967
Car and Driver on Corvette 1968-1977
Car and Driver on Corvette 1978-1982
Car and Driver on Corvette 1983-1988
Car and Driver on Datsun Z 1600 & 2000 1966-1984
Car and Driver on Ferrari 1955-1962
Car and Driver on Ferrari 1963-1975
Car and Driver on Ferrari 1976-1983
Car and Driver on Mopar 1956-1967
Car and Driver on Mopar 1968-1975
Car and Driver on Mustang 1964-1972
Car and Driver on Pontiac 1961-1975
Car and Driver on Porsche 1955-1962
Car and Driver on Porsche 1963-1970
Car and Driver on Porsche 1970-1976
Car and Driver on Porsche 1977-1981
Car and Driver on Porsche 1982-1986
Car and Driver on Saab 1956-1985
Car and Driver on Volvo 1955-1986

BROOKLANDS PRACTICAL CLASSICS SERIES

PC on Austin A40 Restoration
PC on Land Rover Restoration
PC on Metalworking in Restoration
PC on Midget/Sprite Restoration
PC on Mini Cooper Restoration
PC on MGB Restoration
PC on Morris Minor Restoration
PC on Sunbeam Rapier Restoration
PC on Triumph Herald/Vitesse
PC on Spitfire Restoration
PC on Beetle Restoration
PC on 1930s Car Restoration

BROOKLANDS HOT ROD 'MUSCLECAR & HI-PO ENGINES' SERIES

Chevy 265 & 283
Chevy 302 & 327
Chevy 348 & 409
Chevy 350 & 400
Chevy 396 & 427
Chevy 454 thru 512
Chrysler Hemi
Chrysler 273, 318, 340 & 360
Chrysler 361, 383, 400, 413, 426, 440
Ford 289, 302, Boss 302 & 351W
Ford 351C & Boss 351
Ford Big Block

BROOKLANDS RESTORATION SERIES

Auto Restoration Tips & Techniques
Basic Bodywork Tips & Techniques
Basic Painting Tips & Techniques
Camaro Restoration Tips & Techniques
Chevrolet High Performance Tips & Techniques
Chevy Engine Swapping Tips & Techniques
Chevy-GMC Pickup Repair
Chrysler Engine Swapping Tips & Techniques
Custom Painting Tips & Techniques
Engine Swapping Tips & Techniques
Ford Pickup Repair
How to Build a Street Rod
Land Rover Restoration Tips & Techniques
Mustang Restoration Tips & Techniques
Performance Tuning - Chevrolets of the '60's
Performance Tuning - Pontiacs of the '60's

BROOKLANDS MILITARY VEHICLES SERIES

Allied Military Vehicles No.1 1942-1945
Allied Military Vehicles No.2 1941-1946
Off Road Jeeps: Civ. & Mil. 1944-1971
US Military Vehicles 1941-1945
Complete WW2 Military Jeep Manual
US Army Military Vehicles WW2-TM9-2800
Land Rovers in Military Service

2533

CONTENTS

BROOKLANDS BOOKS

ACKNOWLEDGEMENTS

Regular readers of Brooklands Books will know that our aim has always been to make available to motoring enthusiasts material already published about their cars which has become hard to find. Even enthusiasts of cars as recent as the M-series BMWs have difficulty in finding the material they want, and so we are pleased to present it in convenient form in this Performance Portfolio.

As usual, we owe a debt of gratitude to the owners of the copyright regarding the material we have reproduced. Our sincere thanks therefore go to the publishers of Autocar, Autocar and Motor, Automobile Magazine, Autosport, Car and Car Conversions, Car and Driver, Car South Africa, Exotic Cars, Fast Lane, Motor, Motor Sport, Motor Trend, Performance Car, Road & Track, Sports Car Graphic, Thoroughbred and Classic Cars and What Car?

R M Clarke

The 1970's are still viewed as perhaps the least inspired decade in the recent history of the motor car - a time when real and threatened legislation conspired together with dramatic increases in fuel prices to reduce the car to its basic function of a transportation tool. It was almost inevitable that the 1980s should see a reaction, and among the cars in the forefront of that reaction were the M-series BMW's

If the 1970's had temporarily made the purpose-built performance car a rarity, manufacturers nevertheless saw ways of offering attractive products by developing high-performance versions of their family saloons. Few succeeded quite as well as BMW with the M-series cars.

In fact, the first M-series BMW bucked the trend in the 1970s. The M1 was a purpose-built high-performance coupé conceived just before the first fuel crisis struck. Its makers bravely decided to go ahead with plans to build it, but the M1 was the only M-series car of its kind: all the subsequent models were carefully engineered high-performance derivatives of volume-production models.

What made and still makes the M-series cars so special was their unique blend of practicality and driving qualities. Their high-quality construction also makes them durable performance cars, and early examples today offer just as much excitement as they did when new. For enthusiasts of those earlier models, and for owners of the latest M-series cars, this book offers valuable insights into the real character of the cars with Motorsport in their names.

James Taylor

BMW'S SUPERCAR

will cost £20,000-plus ve a family resemblance to the Turbo show car to beat all-comers. At least that's the plan

Artist's impression of BMW's mid-engined Group 5 car. Built by Lamborghini, the car will bear an external similarity to the BMW Turbo show car built in 1972, and, in road-going form, go onto the market with a £22,500 price tag.

1976 HASN'T BEEN, and 1977 won't be, great years for BMW Motorsport GmbH, the young company whose sole purpose it is to guide BMW through the convolutions of international motor sport. Ah, but 1978! The BMW sports directors smile wistfully at the thought, murmur something about mid-engines, and set about persuading you that formula 1 isn't necessarily the most important game in the world.

Jochen Neerpasch, overall Director of this now-self-contained motor sport division (until last year, engine preparation took place in a separate shop) likes to talk about pyramids. There is the formula 1 pyramid, he'll say, in which the World Champion driver forms the apex. There is another — the Manufacturers' pyramid, based from 1977 onwards on the Group 5 Special Production Cars — with a

car at the apex. It is the second structure which interests BMW. Formula 1, Neerpasch thinks, places too much emphasis on drivers. BMW could have decided to build a formula 1 engine when the company's future sporting policy was established early last year, and the World Champion could quite conceivably have been at the wheel of a BMW-powered car within a year or two. But he wouldn't be driving a BMW chassis, because the death of formula 2 BMW driver Gerhard Mitter at the Nürburging in 1969 had evoked a company decision never again to build single-seater racing cars. But there *was* the possibility of the formula 1 engine . . . .

Heady discussions finally turned the long-term involvement towards the World Championship for Makes. For years, BMW had been pushing the CSI for a change in the regulations, and when the organizing body finally relented and drew up the Special Production Car Group 5 for 1975, Neerpasch and others took notice. It wasn't exactly what they were after — there was still tremendous emphasis on turbocharging, for instance — but basically it was sufficient to sway the BMW heirachy, to renew an enthusiasm for racing that had taken a down-swing after the terrific BMW-versus-Ford Group 2 years of 1973-4. That conflict had come to an end with the economic

recession; the next would be curtailed by nothing less than a BMW-dominated pyramid.

The backbone of the Group 5 regulations is a requirement that the racing cars be derived from a production model of which at least 400 units have been built. Neerpasch was adamant that a competitive Group 5 category would only favour mid-engined cars. He knew that BMW had no mid-engined car on the production line — and so he began the long, painstaking business of selling a mid-engined production car to the Board, along with the fringe benefit of a BMW that would do justice to the World Championship for Makes. Surveys were produced, and Neerpasch reasoned that the market was now ripe for a high-cost, high-performance mid-engined two-seater with a "racing" image. Five years ago, the argument went, a mere "sports" image was sufficient. Now, the so-called "sports" BMWs were softer sprung, and more luxuriously-equipped. Consequently, there was a need for a new sports market — a new car on which to stick BMW competition stripes. By the Spring of 1975, the decision had been taken. The mid-engined Turbo was resuscitated.

The Turbo was a one-off Show car produced in 1972 to celebrate the Munich Olympics. A compact (13 ft 9 in.) low-line two seater coupé of very purposeful appearance and sensible engineering (including regenerative crush zones front and rear), it featured big gull-wing doors and very futuristic instrumentation which was even intended to include a radar-signalled stopping distance indicator. A 2-litre turbocharged, fuel-injected, four-cylinder 2002 engine was mounted transversely, in line with its transmission. The engine, with a claimed 280 bhp, was a direct carry-over from those developed for the 2002s in saloon car racing in 1969. Information provided at the time of its presentation described the Turbo thus: "The BMW Turbo gives us a glimpse of what BMW styling will be like in the next generation of cars . . . . it is designed to pave the way for series production . . . . the BMW Turbo shows what the future holds in store."

The words are prophetic, for although the Turbo isn't to form the basis of the new mid-engined BMW, it will have some bearing on the car's exterior. That is a nice development thread from BMW's point of view. For the rest of the car will be the property of Lamborghini, with BMW engineers — notably Motorsport's Paul Rosche and Josef Schnitzer — the guiding hands. Why Lamborghini? Every possible German production facility was approached, but in every case there was no chance of beginning mid-engined work immediately. Lamborghini could, and the Italian firm already had considerable experience with just such cars. Except that BMW want theirs to be "better than a Countach". Ultimately, BMW expect the production car to cost in the region of £22,500,

Above: Formerly a top Porsche long-distance driver, a Ford Competitions Manager, and now Director of BMW Motorsport — Jochen Neerpasch

Right: The turbocharged six-cylinder BMW engine as fitted to the CSL. This 750 bhp power unit could be used in the mid-engined BMW-Lamborghini

straight-six. The final choice will, in effect, be made by the CSI, who are talking about changing the turbocharger boost factor for 1977.

The Group 5 rules impose weight handicaps according to effective engine size, and if the turbocharger factor is increased from 1.4 to 2.0 (as it looks likely to be), then 2.8-litre engines such as the Porsche Turbo's will become less competitive. In that case, BMW will opt for the normally-aspirated engines

1973 to 1975, and this year, the first for the new regulation which admits free-design engines of up to six cylinders, the production-based BMW four has remained a match for the Renault V6. Saloon car interests switched to the USA when Ford withdrew from the battle in Europe at the end of 1974. The Group 2-based BMW CSLs were entered in 13 races in the IMSA series throughout America. They won five of them, finished second in six of them, third in the remaining two, and set six new lap records. More important, the well-oiled Neerpasch race management impressed the American racing scene even more than did the results. BMW Motorsport GmbH claim no immediate responsibility for the uplift — but BMW sales in the United States had increased dramatically by the end of the year . . .

BMW have also been dabbling in Group 5 racing with a number of

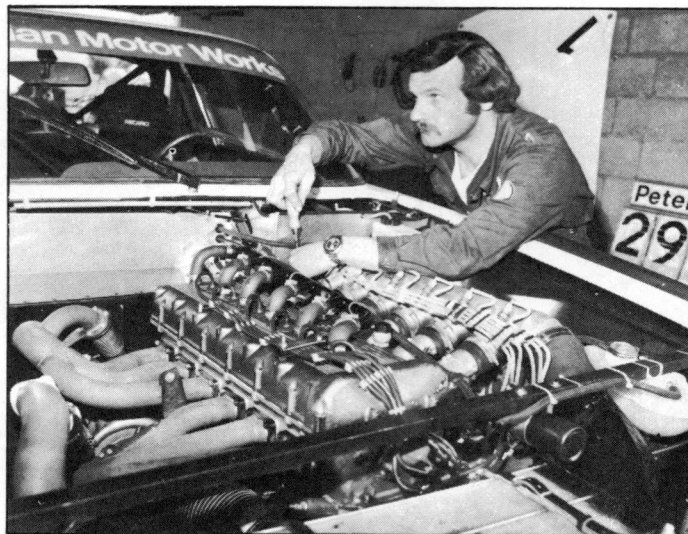

semi-works IMSA-style CSLs, and also developing a fearsome turbocharged version of the same car. British enthusiasts saw the turbo-CSL's debut at Silverstone in May, and it subsequently went to Le Mans where its straight line speed impressed even the Group 6 prototype drivers. The turbocharged engine is reliable now but there are problems with a too-weak drive line and the behaviour of a body/chassis designed for 250 bhp and being asked to cope with up to 800 . . . But this turbo project has served its purpose: if the CSI regulations remain unchanged and demand turbochargers to stay competitive, then BMW have an engine ready.

If the boost factor does change next season, BMW will further develop the 320 saloon that Bo Emmanuelsson has been running in Sweden this year with some degree of success. This would put them in good stead for class rather than outright wins, though there is always the possibility of resurrecting some of those 2-litre turbocharged engines they used back in 1969 . . . . In any case, the production CSL has now been replaced by

the 633 coupé and the new model is unsuited for competition work because of its weight.

BMW's other exciting development — anti-lock brakes — are not being used on the current racing cars. Except for the turbocharged CSL, none of this year's racing cars have been full works entries, and the ATE anti-lock system described in *Autocar* is an expensive one for private teams, at £560 per wheel. Besides, it took works driver Hans Stuck many racing and testing miles to begin to show an advantage with it. The electronics still cause problems with spasmodic operation, and Stuck getting the best from the system was Stuck being able to "feel" exactly when the anti-lock device was going to operate at less than 100 per cent . . .

Development of this and other projects continues, but the real excitement will come with the mid-engined supercar in 1978, a car which for the very best reasons of competition promises to produce a road machine of the extravagant kind that many thought would not be seen again. Work on this progresses apace at Lamborghini, but BMW Motorsport GmbH, a selection of grey single-storey buildings a mile or so away from the tinted-glass air-conditioned headquarters of BMW Munich, is thriving. There is the 3-series programme, a few customer CSLs that will be running in Group 5 next year, the formula 2 engine shop (divided into three parts, so they say — customer engines, works engines, and Stuck's engines. . . .), the advance engineering department, and the merchandising division. The latter has exceeded wildest expectations, and sells team jackets in large numbers all over the world. But in keeping with BMW philosophy, this division also retains an element of exclusivity: at all times and in all places, Jochen Neerpasch must be seen in a jacket that cannot be bought by the public. His road cars have to have the same exclusive touch, so watch out for the ultimate in mid-engined road cars in two years time.

Swede Bo Emmanuelsson has been running a BMW formula 2-engined 3-series car in local races this year. It may form the basis of BMW's 1977 racing plans

and to be identified by a badge saying "BMW — styled by Lamborghini". Construction of the monocoque chassis is beginning now, with one of its features being that it will be able to take any number of different-configuration engines: a decision concerning *which* engine will come from Motorsport GmbH. At present, they are talking about a flat-eight, a flat-10, or a version of the existing

—which they prefer to turbocharged units anyway — for both the Lamborghini-BMW and next year's Group 5 cars.

So the rule change will have an effect on BMW's immediate as well as their long-term plans. Like those of 1975-6, the 1977 season will be nothing more than a year in which BMW's racing programme ticks over. The four cylinder BMW engine dominated formula 2 from

6

BMW M1

Make no mistake—BMW's M1 is more than just a great exoticar. Although it's still recovering from some serious preproduction setbacks, the M1 has required barely two years to establish a reputation as perhaps the finest high-buck GT in the world today. And why not? The sleek, mid-engined production racer was conceived and developed under the basic premise that it would serve as designated captain of BMW's motorsports team during a major portion of this decade. Because of this preordained role, the M1's design incorporated copious amounts of sophisticated technology. This fact immediately manifests itself in no uncertain terms.

Bayerische Motoren Werke has exhibited a longstanding interest in producing high-performance automobiles. The M1 is merely a logical extension of that philosophy. Just one glance tells you that it was born to race. Delve under its skin, and you'll find out how true to this ideal the street version of the M1 remains. Dead stock, it will quickly find its way to the head of a pack of lesser Group 4 cars. In full race trim, BMW expects the M1 to dominate both the Group 4 and Group 5 classes convincingly. They could well be right.

Part of the reason for developing the M1 was to create a corporate PR tool of the first magnitude. The Pro-Car series isn't just a happy coincidence, folks. Eventually, BMW hopes to almost break even on the limited number of street M1s it builds and sells for $65,000 per copy. But the real bottom line is that these cars exist primarily to satisfy the FIA's vehicle homologation requirements. BMW used technology first proven on a number of its other successful racing efforts to make the M1 the car it is, and transforming it into a streetable piece falls strictly into the realm of technical detuning.

While formal responsibility for the M1 project was placed in the hands of BMW Motorsport GmbH, the competition wing of the parent BMW AG, building production versions of the car has become a joint international effort, involving some of the finest craftsmen in Europe. The process begins in Italy. Marchesi builds the tubular space frame there, and Transformazione Italiana Resina makes up the hand-formed fiberglass body. These subassemblies are then bonded and riveted together at Ital Design, where the final bodywork and painting are also performed. Each half-finished M1 is then sent to Baur in Stuttgart, where the suspension, trim, interior and BMW-supplied powertrain are installed. The last stop is at Motorsport headquarters in Munich, where final testing and tuning take place. When it leaves the premises, the M1 is ready to go.

And go it does. The M1 doesn't really come alive until it attains triple-digit cruising speeds. As the velocity increases, its collection of race-bred componentry assumes a dynamic set that is truly synergistic. The M1 can handle day-to-day street prowling without complaint, but it never leaves one with any doubt as to its real *raison d'etre*.

The M1's space frame serves as a heavy-duty backbone, providing a high degree of strength and torsional rigidity. The energy-absorbing front section has been subjected to extensive crash testing, and a full rollcage surrounds the passenger compartment to provide an additional measure of safety.

Nestled just behind this compartment is the heart of the M1: a 3.5-liter DOHC powerplant that serves as another first-class example of Bavarian workmanship. The inline six is mounted longitudinally and uses a special 4-valve crossflow cylinder head to increase efficiency. The high-performance design, which also incorporates hemispherical combustion chambers, was first seen on BMW's championship Formula Two engines. This aluminum alloy head is bolted to a standard 635 cylinder block. The engine's big-bore/short-stroke (93.4 x 84mm) configuration is good for 277 horsepower and 237 pounds-feet of torque.

Feeding the powerplant is a special Kugelfischer-Bosch mechanical fuel-injection system. The indirect port injection is assisted by a supplementary enrichment system for cold starts. A flywheel-fired Marelli solid-state ignition is used to light the fuel/air mix. Other racing influences can be seen in the extensive use of alloy components, seven main bearings supporting a nitrided, forged alloy crankshaft, dry-sump lubrication, a tuned exhaust system and a pair of thermostatically controlled cooling fans that circulate air through the two cross-flow radiators located in the M1's nose.

The powerplant is attached to a ZF 5-speed transaxle via a hydraulically activated Fichtel & Sachs 2-plate dry clutch. The unit has a 4.22:1 final-drive ratio and incorporates a 40% limited-slip differential. For economical high-speed cruising, the top gear is overdriven .704:1.

The M1's fully independent suspension also makes use of lightweight materials and well-proven designs to deliver a ride that offers maximum amounts of comfort and control. Concentric-coil MacPherson struts with Bilstein gas shock ab-

sorbers grace each corner. Up front, unequal-length A-arms work in combination with alloy wheel carriers. Rear suspension is of a similar double-track design, with unequal-length wishbones. Body roll is kept to a minimum by utilizing 23mm front and 19mm rear sway bars.

The M1 relies on 4-wheel vented-power disc brakes to supply adequate stopping power. It would be difficult to heap too many superlatives on these units. They are assisted by the jointly developed BMW-Bosch Anti Brake Locking System (ABS). First announced in 1972 and later track tested on their 3.0 CSL Group 5 car, the ABS uses computerized sensors to activate the hydraulic/electric controls that prevent the wheels from locking, no matter how hard the brakes are applied. Fade also proved to be a virtually non-existent commodity.

We got a chance to find out how good the brakes really were while rolling down the E6 toward Munich at about 150 mph. Some joker in a Citroen traveling considerably slower decided to pull out in front of us. The M1 responded to this potential disaster brilliantly, performing the sudden slow-down quickly, calmly and straight as an arrow. Along with a solid pedal feel and good modulation, BMW has also built into the M1 a 30% anti-dive compensator. This feature was very noticeable and served to create a further feeling of security and well-being every time serious demands were made on the brakes.

The M1's slippery bodywork was designed by Giorgetto Giugiaro, head of Ital Design. Giugiaro's fame can be likened to that of the M1: If he's not the best, he's plenty close. One would expect the work turned out by his organization to be top-notch, and the M1 certainly lived up to our expectations. Because it was planned as a production-based race car, aerodynamic considerations were always paramount. Every line, every vent, every louver in an M1 is there for a reason. The car's .4 drag coefficient allows it to slice through the air with barely the slightest touch of wind noise. The most audible sound comes from air rushing past the oversized, but quite functional, outside mirrors. Popping the headlights up while at speed will also add plenty of secondary harmony to the basic tune.

While the M1's interior would hardly be described as opulent, it is unquestionably a most efficient command post from which one can control the destiny of 2867 pounds of pure-bred excitement. The heavy competition influence becomes readily apparent as soon as one steps down and into the low-slung cockpit. A large white-on-black speedometer and tachometer dominate the central instrument cluster, which houses a full array of gauges. A leather-covered 3-spoke Motorsport wheel holds court over the dash. It has a fore/aft adjustment that can handle the needs of most drivers up to 6 feet. Anyone much taller is going to find things pretty close inside an M1. The low roofline and a rear bulkhead that limits seat travel make head and leg room premium commodities for those of the loftier persuasion.

The ideal placement of the shifter and pedals is further indication of the M1's breeding. But little items like a Becker Mexico cassette stereo, fully adjustable Recaros and power windows provide a welcomed and entirely in-character touch of gentility, which adds to the M1's overall appeal. Ventilation is also in keeping with the BMW approach to things. It's a split system that uses a pair of individually controlled outlets on either side of the dash to allow both hot and cool air to be circulated simultaneously.

On the road, the M1 is a match for anything. Period. Its ultra-low center of gravity, excellent balance and weight distribution, superb suspension and Pirelli P7 tires combine to give the M1 handling capabilities that far exceed the skills of most who drive it. The M1 was designed to handle upwards of 800 horsepower. With just about a third of that potential realized in the street version, the car generally felt like it was running on rails.

Straight-line acceleration is pretty much on a par with most of the heavy hitters in the GT world. As the going gets twistier, the M1's handling edge becomes more pronounced. The decision was made to go with a direct-acting rack-and-pinion steering gear because BMW engineers felt that, in this kind of car, a driver needs as much sensory input as possible. To help offset the increased steering effort, they dialed a good deal of caster into the front-end geometry. This results in directional control that is positive at all speeds but that doesn't require an inordinate amount of muscle power.

While driving an M1 in town is almost certain to be a frustrating experience for anyone behind the wheel, the car itself is willing to negotiate most city situations with unexpected aplomb and good cheer. The engine's broad power band allows it to pull in any gear from under 2000 rpm right up to the 7000-rpm redline. The shift linkage is always crisp and precise, and the only real drawbacks worth mentioning are that the car's shovel nose, large turning circle and somewhat restricted side and rear vision make maneuvering in and out of tight spaces a bit of a chore.

Once out in the open, it's a different story. While the engine is hardly "cammy," it really comes on strongest as the tachometer slips past 4000 rpm. Like several of its exotic counterparts, the M1 has a rev limiter to prevent one from indulging in too much of a good thing. Although there are usually some sonic clues as to the engine's specific activities at any given moment, the interior noise levels always remain quite acceptable, even when the roadside scenery is passing by in triple time.

Beyond 150 mph, the front end begins to feel marginally lighter. Our only real white-knuckle moments came at the top end of an all-out speed run. The M1 had just passed its 162-mph factory-spec top speed and was still accelerating when we ran into some fairly heady cross-wind buffeting. With visions flashing before us of T.C. Browne standing atop his inverted Miura, we decided to let discretion rule the day and backed down gracefully.

Our time in the M1 was short, but it was still possible to see that BMW had done what they set out to: build a road-going racer *par excellence*. Because of the costs involved in federalizing, as well as a domestic demand that far outstrips current production rates, they have no plans to export their street racer to our shores. Only a few privately imported examples have made the journey thus far.

If you just can't visit the M1 on its home turf, the next best place to see BMW's *wundercar* is at an IMSA GT race. While most of the M1s in competition will be March-prepped monocoque monster cars—distant cousins to a production M1, to be sure—several may be showing up in near Pro-Car trim. In either case, it promises to be an interesting show and your chance to see part of a legend in the making. **SCg**

SPECIFICATIONS

BMW M1

GENERAL

Vehicle type	Mid-engine, rear-drive, 2-pass. GT
Base price	$65,000 (113,000DM)
Options on test car	None
Price as tested	$65,000

ENGINE

Type	Inline six, water cooled, alloy head and block, 4-valve hemispherical cylinder head
Bore & stroke	93.4 x 84 mm
Displacement	3453 cc
Compression ratio	9.0:1
Fuel system	Kugelfischer-Bosch mechanical injection
Recommended fuel	Premium leaded
Emission control	European
Valve gear	Double overhead camshafts, chain driven
Horsepower (SAE net)	277 at 6500 rpm
Torque (lb.-ft., SAE net)	239 at 5000 rpm
Power-to-weight ratio	10.35:1 lb./hp

DRIVETRAIN

Transmission	ZF 5-speed manual w/overdrive, 40% limited slip
Final drive ratio	4.22:1

DIMENSIONS

Wheelbase	100.8 in.
Track, F/R	61.0/62.0 in.
Length	171.7 in.
Width	71.8 in.
Height	44.9 in.
Ground clearance	4.9 in.
Curb weight	2867 lb.

Weight distribution, F/R	1262/1605 lb.

CAPACITIES

Fuel capacity	27 gals.
Crankcase	9.6 qts.
Cooling system	24 qts.

SUSPENSION

Front	Independent, MacPherson struts w/concentric coils and Bilstein gas shocks, double unequal-length control arms, sway bar
Rear	Independent, MacPherson struts w/concentric coils and Bilstein gas shocks, double unequal-length control arms, sway bar

STEERING

Type	Rack and pinion, direct acting
Turns lock-to-lock	3
Turning circle, curb-to-curb	42.7 ft.

BRAKES

Front	11.8-in. vented discs, power assist
Rear	11.7-in. vented discs, power assist

WHEELS AND TIRES

Wheel size	7 x 16 in./8 x 16 in.
Wheel type	Cast alloy
Tire make and size	Pirelli P7: 205/55 VR 16 front 225/55 VR 16 rear
Tire type	Steel-belted radial

9

BMW M1

Illustration prepared by BMW AG, reprinted courtesy *Autocar* Magazine

1. Hydraulic reservoir
2. Brake servo
3. Ventilation duct
4. Washer reservoir
5. Battery
6. Fuse box
7. Pirelli P7 205/55 VR 16 tires
8. Disc brakes
9. Foot rest
10. Hood release
11. Cockpit air vent
12. Fuel tanks
13. Air-conditioning pump
14. Engine-cooling inlet
15. Kugelfischer-Bosch fuel injection
16. Header tank
17. Electronic ignition
18. 3.5-liter DOHC engine
19. Pirelli P7 225/50 VR 16 tires
20. Disc brakes
21. ZF 5-speed transaxle
22. Mini-spare
23. Oil tank filler

BMW M1

An exoticar that's more racer than reality

by Jerry Sloniger

JERRY SLONIGER

DRIVING Impression

Once BMW decided to build a mid-engined image-maker, they persisted in building a race car first and then detuning it for the road. They figured it was easier that way, and I guess they're right, if you ignore the inevitable stretch in development time for track perfection. Of course, such practice assumes that every M1 buyer will appreciate the ultimate in fine-tuned suspensions rather than image consciousness.

At least we now have a chart of the triple play which will bring you the M1s—all 400 of them. BMW wheedled motor racing authorities into accepting them for Group 4 competition in 1979—before the required 400 vehicles (homologation number) were completed—on the strength of a promise to make them by year's end. All systems are now on red alert.

It all begins with Marchetti, a Turin tube-bender known for open wheelers and some Lamborghinis. He sends the frames on to designer Giugiaro (Ital Design) who based the final design on a one-time BMW turbo show car.

Ital Design fits the vehicle's fiberglass panels, seats and some trim before forwarding to Baur of Stuttgart, where engines, transaxles, suspensions, dashboards, etc., are given a final fit.

The engines come off a quasi-assembly line at the regular BMW plant, since the block is shared with the 635 coupe. For the factory team, of course, BMW Motorsport will prep the turbocharged Gp 5 race engines, while Ron Dennis in England and Osella of Italy will take care of the Gp 4 customers.

The first 35 cars will be Gp 4/Procar machines for a special IROC-like series pitting the fastest five qualifiers at each European Formula One race against 15 hopefuls. Twenty of their 30 allocated racers are already ordered. Thereafter, road versions will be launched at $53,000 a copy.

Our test vehicle was a preproduction road version, which most people compared to, or mistook for, an Italian exotic. Interesting, since BMW is vigorously promoting it as the fastest production car built in Germany. Which brings us to two rather important considerations: Is the M1 too Giugiaro-Italian to have Bavarian identity? And, did all those GP-caliber test drivers bias its track manners too much towards the people who could conceivably afford one?

BMW's engine may have only 6 cylinders, but it boasts 24 valves operated by two overhead camshafts. It started immediately, hot or cold, and idled gently, but a maximum torque at 5000 rpm told the

tale; it was very peaky for daily use. Modest turbocharging across the line might have given the M1 more of a universal engine, but BMW was burned once by the 2002 Turbo and opted instead for the more classic, highly tuned approach. Such engines do best with many shifts, but the factory's 5-speed ZF box had a lever strongly biased to 2nd and 3rd, so we had to watch downshifts from 5th because it was easy to have mistakenly plopped it into 2nd. Alternatively, it was easy to override the resistance which protected 1st and get into the separate plane for reverse.

A sticky throttle linkage was certainly unique to our hard-used development car, but its 2-plate clutch was quite normal. Pedal movement was very long, with only enough grip at the bottom for initial creep, whereas complete engagement came in at the top ½ inch of pedal trav-

el, where it bit like a racer.

Once rolling, the M1 became more impressive in direct ratio to increasing speed. The first day of winter was not ideal for independent measurements, but 155 mph was easily reached before traffic intervened. So, the M1's factory-claimed 162-mph top speed seemed attainable, as did its sub-6-second 0 to 60 time, or less than 21 seconds to 125 mph. The clue came from an eerie situation where we played around the 100-mph mark only to look down and find ourselves doing 125 mph. It touched 100 in 3rd, 140 in 4th and stopped so surely we could virtually ignore the inner-vented discs.

Around town the P7 Pirellis registered each cobblestone individually, but they also turned frost-rumpled highways into billiard tables (although BMW is not alone in finding imbalance problems above 150 mph with them). Dual-wishbone suspension and Bilstein shocks were fine-tuned for typical mid-engine handling: neutral to a very small border realm, then oversteer. However, packed snow discouraged attempts to best BMW's claim of near-l-g lateral forces. The small thick-rim wheel transmitted every input with such precision and feel that we forgot there wasn't any servo.

Yet a weight distribution of 44/56% eliminated the car's feeling of heaviness around town. It was least nervous when driven fast.

Regarding the M1's interior, tweedy seats give fine lateral support, and there is surprising elbow room, but fore-aft adjustment is scant, aided for shorter drivers by a wheel which slides in or out. Restricted headroom is partly due to sitting more upright than usual, but that is necessary in order for one's thighs to fit. Somehow BMW did find space in a very narrow footwell for a left-foot brace. Brake and throttle are ideally placed so that you can easily roll your right foot either way without having to lift it, in best formula driving manner.

Tach and speedo (lit in white, not the Bavarian red) are very visible, but your hands hide the four smaller dials. Switch gear comes from BMW's 7-series, heater controls from the 3-range, and general ergonomics show the work of a large-volume producer, a real plus in the exotic field. There are air vents all over the dash and ample heat for sub-freezing days, but that expanse of glass suggests optional air in desert regions. No heat seeps in from the engine, and even the "luggage" bin behind it (and above the gearbox) does not get overheated. The car swallows the equivalent of a two-suiter, a couple of overnight cases and two or three soft kits plus spare wheel. Two thin door pockets and two small bins (one locking) serve the cockpit.

BMW claims production models will be far quieter than the prototypes, a good move since general noise level becomes intrusive at high speed, and wind noise is annoying despite wind tunnel work to achieve a drag coefficient of "under 0.4."

Vision is the inherent mid-engined problem. A backbone chassis allows low door sills, for ease of entrance, and large side windows. The nose falls right out of sight. Vision to front or sides is excellent—and even better with the pop-up headlights in place. However, the (heated) rear window is a slit soon covered by road film, and the thick rear quarter panels defy X-ray vision, so that merging from an on-ramp is tricky at best. Fortunately, both outside mirrors adjust electrically, from one switch, and windows are electric too.

The car doesn't seem to spread around you in all directions like some exotics, but you still want to avoid cramped parking places. As a near-43-foot turning circle suggests, you don't park an M1—you dock it.

The real significance of this car is that BMW has emerged from oil-shock to reclaim its sports car/swinger image. The moot question is whether enough people, sure to drool over one, will have not only the price but the ambition to learn how to drive one well. There certainly is more macho here than in a Turbo Porsche, so BMW will probably sell the first 400 easily enough, but the additional bonus few thousand sales may be lost to the M1's specialization.

MT

SPECIFICATIONS

ENGINE

Type	In-line six, dual OHC, dry-sump lubrication
Bore & stroke	84.0 x 93.4 mm
Displacement	3452 cc
Compression ratio	9.0:1
Fuel system	Mechanical Kugelfischer injection
Recommended octane number	N.A.
Valve gear	Dual OHC, 4 valves per cylinder
Horsepower (SAE net)	277 at 6500 rpm (7000-rpm redline)
Torque (SAE net)	239 lb.-ft. at 5000
Power to weight ratio	10.3 lb./hp (empty)

DRIVETRAIN

Transmission	5-speed ZF, integral w/differential
Final drive ratio	4.22:1 (40% limited-slip)

DIMENSIONS

Wheelbase	100.9 in.
Track, F/R	61.1/62.1 in.
Length	171.8 in.
Width	71.9 in.
Height	44.9 in.
Ground clearance	4.9 in.
Empty weight	2865 lb.

SUSPENSION

Front	Independent, dual wishbones w/alloy hubs, Bilstein shocks, 0.9-in. stabilizer
Rear	Independent, dual wishbones w/alloy hubs, Bilstein shocks, 0.75-in. stabilizer

STEERING

Type	Rack and pinion
Turns lock-to-lock	N.A.
Turning circle, curb-to-curb	42 ft., 8 in.

BRAKES

Front	Inner-vented discs w/fixed saddles, servo & balance
Rear	Inner-vented discs w/fixed saddles, servo & balance

WHEELS AND TIRES

Wheel size	7 x 16 front/8 x 16 rear
Wheel type	Cast alloy
Tire make and size	Pirelli P7, 205/55 VR 16 front/225/50 VR 16 rear
Tire type	Radial
Recommended pressure, F/R	N.A.

ACCELERATION (WORKS FIGURES)

0-100 kph (62 mph)	5.6 secs.
0-200 kph (125 mph)	20.7 secs.
Top speed	162 mph

M1 - the mighty one

BMW's M1 Coupé, styled by Giugiaro and built by Bauer (originally it
was to be built by Lamborghini, before that company found itself in
financial difficulties), is now on sale as a road car, to challenge the
Porches and Ferraris. You'll have to wait for right-hand drive versions,
but we went out to Germany to put this new supercar through its paces

Report by Jeremy Sinek. Photographs by Maurice Rowe

In spite of well shaped seats and reach-adjustable wheel, the driving position is nothing special: seat travel is generous but the rear bulkhead limits recline

Conservative interior treatment, but neatly and tidily executed with well-fitting panels, and good quality carpet

The comprehensive instrument display is not up to BMW's usual high standards of clarity and sensible location

In spite of the bulk of the space-saver rear tyre (illegal in the UK), there's fair boot space for 'squashable' luggage

The front of the car, however, is fully occupied by the radiator, battery, fuse box and other assorted ancillaries

THERE EXIST among car makers an elite few, the very mention of whose names is guaranteed to stir the imagination of even the most unaware of motorists, and indeed, even among those who have no part in the world of motoring, names like Ferrari, Porsche, Maserati and Jaguar are recognised as being synonymous with the ultimate in automotive speed and style.

It is not a status that is casually attained, for it is founded on a history of sporting honours earned in the purest and most elevated forms of motoring competition. In many cases, the reputations were first earned by triumphant Le Mans or Grand Prix racers, and only subsequently by the road cars that capitalised on engineering strengths forged in the heat of competition and basked in the reflected glory of their racing stablemates.

To date, BMW as we know it since it rose from the ashes of near-bankruptcy a couple of decades ago could never conceivably have qualified for membership of that elite band. Its products, however excellent, have lacked the pedigree, tradition or style for its name to become a household word. BMW's has been the role of the upstart, the infant prodigy among car-makers. And its sporting reputation has been founded on racing modified saloon cars: recognised and respected by afficionados but lacking the wider familiarity that can only be earned by success in more widely publicised forms of competition such as Grand Prix racing or Le Mans.

But now BMW has a car with which it could make the transition from whizz-kid to senior partner in the hierarchy of reputations, and the World Championship of Makes will be the field of endeavour where it must succeed or fail.

So it was to this end, BMW are at pains to point out, that the M1 was first and foremost conceived: as a purpose-designed racer, equipped with an engine ultimately able to put out over 800 bhp in turbocharged Group 5 form, and with an uncompromising suspension and braking specification to match the potency of such an engine.

Yet, although over two years have passed since the project was first seen in public, the M1 has yet to make its racing debut (other than in incestuous competition with itself in last year's series of one-make Procar races at European Grand Prix venues), for to be eligible to compete in Group 5, a racer must be derived from a car that has (in this case) Group 4 homologation, which in turn requires a minimum production of 400 units in 24 months, a target which so far has only been half realised. No doubt the abortive link-up with Lamborghini — who were contracted to assemble the cars but whose failure to deliver led BMW to take the work away from the financially floundering Italian firm — has delayed the flow of completed production cars and hence the commencement of competition in earnest.

In the meantime, then, BMW is in the position of building and marketing a svelte road-going sports car in the finest supercar traditions, stunningly fast and yet with reserves of dynamic ability far in excess of its engine's output.

At the heart of this formidable machine is a massive steel frame of square section tubing which incorpo-
continued over

15

rates an energy absorbing front section and full roll-over protection. To this is bonded and riveted a glass-fibre shell — from the pen of Italdesign and claimed to be designed for function before style — which clothes mechanical ingredients of impeccable conception. At each corner the suspension is by unequal length wishbones, with height-adjustable coil springs concentric with Bilstein gas-filled damper units, and an anti-roll bar at each end. The steering is by rack and pinion, braking by massive ventilated discs mounted outboard at each end on light alloy hub carriers, and the 16 in diameter alloy wheels are shod with ultra low-profile P7 Pirellis: 205/55 VR 16 on 7 in rims at the front, 225/50 VR 16 on 8 in rims at the rear.

Drive is transmitted to the rear wheels through a five-speed ZF gearbox, with an integral final drive incorporating a 40 per cent slip limited-slip differential. And mounted in line with this transaxle, behind the cockpit and ahead of the rear wheels, is an engine which has little left in common with the 635 CSi straight-six unit on which it is based.

It still has a capacity of 3453 cc achieved from 93.4/84.0 mm bore/stroke dimensions, and a seven-bearing crankshaft running in a cast-iron block, but in the M1 it is mounted vertically, not at a slant as in its original application. In place of the conventional sump is a fully pressurised dry sump lubrication system, while the block is topped by a light alloy cylinder head featuring four valves per cylinder and twin chain-driven overhead camshafts. On a compression ratio of 9.0:1 it burns 98 octane fuel supplied by Bosch-Kugelfischer mechanical injection and ignited by fully-electronic Marelli ignition with contactless regulation from the flywheel.

In naturally-aspirated Group 4 trim this classic engine is capable of producing 470 bhp (DIN) at 9,000 rpm, but in the road car it develops a more modest, but nonetheless highly potent, 277 bhp (DIN) at 6,500 rpm and a maximum torque of 239 lb ft at 5,000 rpm.

Detuned it may be by comparison with the racing version, but with a specific output of just over 80 bhp per litre, the road car's engine is no soft and lazy touring engine: it's a thrusting, energetic go-getter, strong for its size and eager to prove it, urging the driver to let it have its head.

Yet for all that, it's also capable of self-restraint and doesn't chafe at the bit in the hands of a driver who opts for gentler progress. In short, it's an engine of rare talent, combining an equable temperament with a range of performance that can only be described as breathtaking.

It starts promptly, hot or cold, and although not all of the six cylinders fire up at once, they do all very soon settle down to a slow, even, reliable idle.

If asked to, the engine will trickle along without protest at 1,000 rpm in any gear, and then pull away cleanly and strongly when you open the throttle. And from then on it surges relentlessly up the rev range, passing through a faint hint of a dip in the power curve at around 3,500 rpm and then taking on a purposeful howl from 4,000 rpm all the way to the red line at 6,700 rpm — or even beyond, to the 6,900 rpm ignition cut-out, if you're not careful. The fourth gear acceleration figures illustrate it perfectly: 20-40 mph in 6.2 sec, 40-60 in 6.0, 60-80 in 6.3, 80-100 mph in 6.0 sec, the power available rising in perfect unison with the power required.

This near perfect power delivery is matched by a flawless choice of gearing which allows the M1 almost exactly to attain its maximum revs in top, to reach a maximum speed that now displaces the Porsche 3.3 Turbo's as the highest that *Motor* has ever measured on a production car. On a dry, still day along a flat, straight autobahn near Munich our M1 test car soared past 150 mph to reach with almost disdainful ease a mean speed of 161.0 mph, equivalent to an engine speed of 6,600 rpm and with the needle of the slightly optimistic tachometer buried in the red at an indicated 6,850 rpm.

Among other supercars only the Turbo Porsche has recorded significantly better standing start acceleration figures than those of the M1, even though the BMW was penalised by a surface that provided a less than ideal degree of traction off the line. Such times as 5.6 sec to 60 mph, 13.1 sec to 100 mph and 20.1 sec to 120 mph (compared to 5.3, 12.3 and 19.1 sec respectively for the Porsche) firmly establish the M1 as ranking among the very quickest of superquick cars.

Moreover, these figures are not just of academic interest: the performance they reflect is fully useable, and the driving technique they require is no hardship. For all its low speed tractability, it's on high speeds that the engine thrives — noisily to be sure, but with an exultant howl and an underlying smoothness that encourages rather than discourages the use of maximum revs in the lower gears. Likewise, in fifth, 130 mph on the autobahn is a comfortable and realistic cruising speed in spite of the uproar from the engine bay.

In part this can be credited to a usefully tall fifth gear that gives 24.3 mph per 1,000 rpm, and which is part of a well chosen set of ratios allowing 48, 71, 101 and 136 mph at 6,700 rpm in the lower ratios. These are selected by a shift pattern that is laid out in the usual ZF manner with the upper four ratios forming an H, and first on a dog-leg to the left and back, with reverse more or less opposite but protected by a detent from accidental selection. With the important proviso that you use the full extent of the clutch's travel, the gearshift works with speed and precision and on the move smooth gearchanges are easily mastered. Paradoxically, though, the M1 is not easy to get smoothly away from rest, for the engagement of drive is quite abrupt, yet the clutch has a woolly feel that tends to mask the point at which it actually occurs — and an accelerator action that is a little stiff and heavy over its initial travel only makes matters worse.

Considering the number of factors conspiring against a good overall mpg figure for our test — not least the shorter than usual test mileage and the fact that much of it occurred on speed limit free German autobahns — the final result of 15.4 mpg is surprisingly good. And in conjunction with a tank capacity of no less than 25.5 gallons it adds a new dimension to the M1's potential as an autobahn cruiser, allowing a range of anything up to 700 miles if BMW's claim of 27.7 mpg at 75 mph is to be believed.

Not all mid-engined sports cars manage to realise fully the handling potential inherent in their layout: some understeer strongly at low speed, some display vicious lift-off oversteer characteristics, others are unstable at speed. None of the accusations can be levelled at the M1. It is agile and well balanced on tight low-speed corners, yet rock steady on sweeping autobahn curves at speeds far in excess of 100 mph. Most of the time talk of understeer or oversteer seems quite superfluous; it simply sails round corners with seemingly limitless reserves of roadholding, guided by steering that approaches perfection in its precision and its weighting. On tight bends it is possible to power the tail out into an easily held slide in first or second gear but otherwise, short of hopelessly misjudging a corner, you feel you'll never approach anywhere near its limits in the dry. In the wet, inevitably, you must exercise due caution, for oversteer can be easily provoked either under power or on lift-off, but even

then you don't need exceptionally sharp reactions to bring the tail to heel in time.

If the M1 has anything approaching a vice, then it is one that is shared by most other cars which are similarly shod with ultra-wide low profile tyres — a tendency to follow cambers and white lines, which demands a conscious correction by the driver. On the M1 this characteristic can give rise to some weaving when braking hard on an uneven surface, but that is our only criticism of an otherwise ideally judged braking system — progressive, powerful and utterly tireless, the four massive ventilated discs repeatedly dragged the M1's speed down from 130 mph with never so much of a hint of wilting under the strain.

As a driving machine, then, the M1 is a masterpiece: were this a formal road test it would so far have earned the maximum five stars each for performance, handling and brakes and four for transmission. But then that much is only to be expected from a road-going development of a purpose-designed racer. The real challenge is of how it measures up as practical, comfortable everyday transport. Is the M1 simply a racer made road-legal, or is it a civilised production car in its own right, able to stand comparison with the best from Ferrari, Porsche, Maserati et al?

The answer has to be a qualified, but nonetheless unhesitating, 'yes', and whether or not the 'but' is relevant depends entirely on the physical dimensions of the potential purchaser. Quite simply, if you're anything much over average height the M1 won't fit you. The low build and the long engine behind the rear seats leave little room to spare in the cockpit for larger people, the too-close rear bulkhead providing them with a Hobson's choice between a comfortably reclined backrest angle but insufficient legroom, or adequate legroom but an upright backrest position that aggravates the lack of headroom. So if you're tall, you're

continued over

M1-THE MIGHTY ONE

committed to an awkward posture for which the basic good shape of the seat itself is little compensation. And even if you're a sensible driver the basic driving position is not ideal, in spite of the reach-adjustable wheel: a height adjustment, allowing you to lift the wheel clear of your thighs, would be more useful, and it's a pity that the left foot-rest so thoughtfully provided is too close to the clutch pedal and gets in the way when you're changing gear. The pedals are heavily offset to the right, too. On the other hand the gear lever is well placed, as are the minor switches. Overall, then, seating comfort is adequate if you are moderately proportioned, but possibly a major flaw if you're not.

Otherwise, it's nearly all good news. The standard of ride comfort is almost miraculous for this breed of car, largely a product of flawless matching of spring and damper rates. It's smooth even at low speeds and over small bumps, and just gets better and better as the speedo needle climbs. An outstanding achievement.

Not that the ride always *sounds* as smooth as it is. Bumps are heard more than they're felt, and cobbles, for example, set up quite a rumpus. On the other hand tyre roar is held quite well in check — far more so than in various similarly shod Porsches we've driven — and wind noise is kept down to inoffensive levels regardless of cruising speed. Thus it's noise from mechanical sources that dominates, a combination of transmission whine and a curious hissing sound from the engine orifices at moderate speeds, and a rising banshee wail from the engine as the revs rise: a sound level reading of 90 dBA at peak revs in second gear isn't the highest we've ever recorded, but it comes close. On the other hand, as noises go it is of the very best kind — an enthusiast's joy — and we wouldn't want it any other way.

The M1's instruments owe less to BMW saloon practice (a few large, clear dials under a single angled pane of glass) than to British tradition, with a comprehensive collection of individual dials (speedo, tacho, oil pressure, oil temperature, water temperature and fuel) each with its own reflective glass cover, some of which are likely to be at last partly obscured by the wheel or the driver's hands.

In other respects the cockpit fittings draw heavily on existing BMW parts bins and the result is a comprehensively appointed interior that clearly identifies the M1 as a 'real' car, and not a glorified kit car; if it's not perfect, neither are any of the opposition.

Thus there is an elaborate heating and ventilation system with four vertical sliders to control it and four face-level fresh air vents, all of which look rather more impressive than the results they produce, for the heater's output is slow to respond to adjustment and the flow through the face-level vents, though independent of the heater, is no more than modest even on full boost. On the other hand, the very fact of their existence is convincing testimony to the M1's seriousness as a production car, and if that isn't enough, the fitting of a full and effective air conditioning system should clinch it.

Other fittings that smack more of 'road' than 'racer' are the l.e.d. digital clock, electric window winders, stereo radio/cassette player with automatic electric aerial, and a pair of superb electrically adjustable door mirrors.

The latter, aided and abetted by an effective wash/wipe system and a clear view directly to the rear through the interior mirror, make for good visibility on the move. Parking, however, is a different story, for the nose drops right out of sight and it's particularly difficult to judge the width of the car from the driver's seat — though fortunately the tendency is to over rather than under-estimate it. Which is just as well, since it would be a shame to let a car-park scrape damage what must be the finest quality glass-fibre panelling to be found on any car. The fit and finish of the panels is immaculate, and the structure has a hewn-from-solid integrity rarely encountered on cars with a separate chassis. Similarly the interior treatment, although fairly low-key and conservatively styled, is neatly assembled with no ragged edges.

The message is clear. While the M1's intended destiny, the very reason for its existence, is ultimately to win races, and hence prestige and glory for its creators, it is also capable in the meantime of fulfilling a very different role. As a road car it demands to be taken seriously. That it is dynamically superb is only to be expected: the real achievement lies elsewhere, in an overall blend of qualities that add up to a very 'complete' car which is subtly developed, beautifully made and lavishly equipped. With the M1, BMW has a serious and tempting alternative to the best that the likes of Porsche and Ferrari can put up against it. And a worthy addition to the ranks of the world's truly great cars.

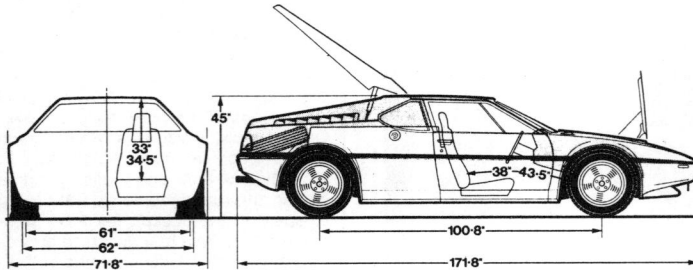

PERFORMANCE

CONDITIONS
Weather	Dry, wind 1-4 mph
Temperature	37°F
Barometer	28.3 in Hg
Surface	Dry tarmacadam

MAXIMUM SPEEDS
	mph	kph
Mean	161.0	259.0
Best	161.2	259.4

Terminal Speeds:
at ¼ mile	102	164

Speed in gears (at 6700 rpm):
1st	48	77
2nd	71	114
3rd	101	162
4th	136	219

ACCELERATION FROM REST
mph	sec	kph	sec
0-30	2.5	0-40	2.1
0-40	3.4	0-60	3.1
0-50	4.5	0-80	4.5
0-60	5.6	0-100	5.9
0-70	7.1	0-120	7.8
0-80	8.8	0-140	10.2
0-90	10.8	0-160	12.9
0-100	13.1	0-180	17.0
0-110	16.3	0-200	21.6
0-120	20.1		
Stand'g ¼	13.8		

ACCELERATION IN TOP
mph	sec	kph	sec
20-40	—	40-60	—
30-50	7.5	60-80	4.8
40-60	7.8	80-100	4.8
50-70	7.9	100-120	5.2
60-80	8.5	120-140	5.5
70-90	8.9	140-160	5.5
80-100	8.9		

ACCELERATION IN 4th
mph	sec	kph	sec
20-40	6.2	40-60	3.7
30-50	6.2	60-80	3.9
40-60	6.0	80-100	3.6
50-70	6.0	100-120	4.0

mph	sec	kph	sec
60-80	6.3	120-140	3.7
70-90	6.1	140-160	3.7
80-100	6.0		

FUEL CONSUMPTION
Touring*	20.0 mpg	14.1 litres/100 km
Overall	15.4 mpg	18.3 litres/100 km
Govt tests†	31.7 mpg (56 mph)	27.7 mpg (75 mph)
Fuel grade	98 octane	5 star rating
Tank capacity	25.5 galls	116 litres
Max range	510 miles	821 km
Test distance	391 miles	630 km

*Estimated consumption midway between 30 mph and maximum less 5 per cent for acceleration †Factory figures.

NOISE
	dBA	Motor rating*
30 mph	67	13
50 mph	72	18
70 mph	76	24
Max revs in 2nd (1st for 3-speed auto)	90	64

*A rating where 1 = 30 dBA and 100 = 96 dBA, and where double the number means double the loudness.

WEIGHT
	cwt	kg
Unladen weight*	25.6	1300
Weight as tested	29.3	1488

*manufacturers figure

Performance tests carried out by Motor's staff in Munich, W. Germany.

Test Data: World copyright reserved; no unauthorised reproduction in whole or part.

GENERAL SPECIFICATION

ENGINE
Cylinders	6 in line
Capacity	3453 cc (210.5 cu in)
Bore/stroke	93.4/84.0 mm (3.71/3.31 in)
Cooling	Water
Block	Cast iron
Head	Alloy
Valves	4 per cylinder, dohc
Cam drive	Duplex chain
Compression	9.0:1
Carburetter	Kugelfischer-Bosch mechanical injection
Bearings	7 main
Max power	277 bhp (DIN) at 6500 rpm
Max torque	239 lb ft (DIN) at 5000 rpm

TRANSMISSION
Type	5-speed manual
Clutch	Twin-plate
Actuation	Hydraulic

Internal ratios and mph/1000 rpm
Top	0.704:1/24.3
4th	0.846:1/20.2
3rd	1.14:1/15.0
2nd	1.61:1/10.6
1st	2.42:1/7.1
Rev	2.86:1
Final drive	4.22:1

BODY/CHASSIS
Construction Glassfibre body bonded and rivetted to box-section steel frame

SUSPENSION
Front	Double wishbones; gas-filled dampers with concentric coil springs, adjustable for height; 23 mm anti-roll bar.
Rear	Double wishbones; gas-filled dampers with concentric coil springs, adjustable for height; 19 mm anti-roll bar.

STEERING
Type	Rack and pinion
Assistance	No

BRAKES
Front	11.8 in ventilated discs
Rear	11.7 in ventilated discs
Park	On rear
Servo	Yes
Circuit	Split
Rear valve	Yes
Adjustment	Automatic

WHEELS/TYRES
Type	Light alloy, 7 × 16 front/8 × 16 rear
Tyres	Pirelli P7 205/55 VR 16 front 225/50 VR 16 rear
Pressures	36/43 psi F/R

ELECTRICAL
Battery	12V, 55 Ah
Earth	Negative
Generator	65A alternator
Headlights type	2 × Rectangular

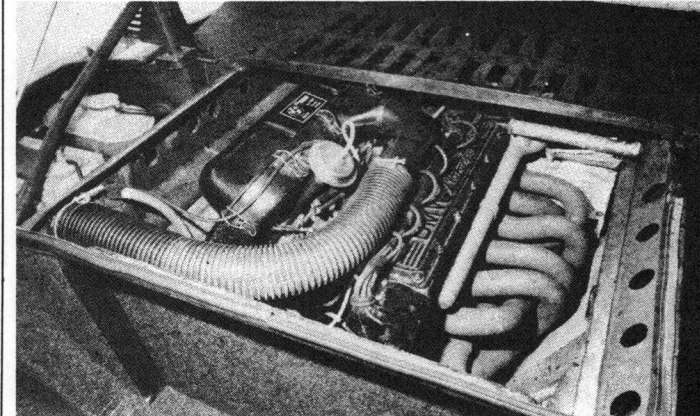

Viewed from the rear, the engine displays its magnificent tubular exhaust manifold

BMW M535i five speed

The fastest—and most economical BMW in production in South Africa is a true luxury car, as well

This is South Africa's fastest production BMW — even faster than the fabled 530 — and one of the most spectacular road cars available to South African motorists.

Introducing the 160 kW 3,5-litre engine of the "7 Series" flagship, with fuel injection and overdrive-style five-speed gearbox, the new M535i (that "M" is for "Motor Sport") is a carefully-balanced compromise between outright performance and practical everyday motoring requirements.

It can black-stripe away from rest, with the limited-slip diff provoking tyre smoke at both rear wheels, and it goes a shade over a true 200 km/h on a level road. Yet it cruises quietly and gently, has full air-conditioning, and can return

fuel economy figures to rival some four-cylinder models.

NEW FEATURES

The M535i is based on the 530 — which is replaced in local manufacture — with the handsome and efficient "5 Series" coachwork, and the kind of handling and ride which has made this range of fine cars famous. While it has little external difference from its sisters in the "5 Series" range, it is a sedan in the sports car class, and given a touch of distinction by big steel-belt radials on wide Mahle light alloy wheels with cross-spoke styling.

The all-independent suspension is uprated on this model, with heavier stabiliser bars front and rear and Bilstein

gas-filled shockabsorbers, and it has a ZF 25 per cent limited-slip differential to give maximum traction.

It has much of the comfort of the top BMW Executive models, with custom air-conditioning, a programmable Pioneer radio/tape deck, central locking system, power windows, and Recaro sports seats. As a final touch, the steering wheel is from the racing M1 coupé.

MUSCULAR MACHINERY

The new big-six 3,5-litre engine made its debut last year in the flagship BMW 735i Automatic (Car Road Test, August 1981) and is remarkable for its willingness and range, with up to 6 200 revs usable. It produces 160 kW at 5 200 revs, and is matched with long-legged gearing in this unusual performance car: a slick four-and-overdrive gearbox with an indirect top.

In the direct-drive fourth the car is geared for 37,8 km/h per 1 000 revs (which already is virtual overgearing) and goes to an amazing 46,4 km/h per 1 000 revs in 5th, which is a true "economy gear".

The clutch has a fairly heavy action, and bites well.

PERFORMANCE

The M535i is a more-refined car, and does not have the fiery character of the earlier 530 (CAR Road Test, January 1978). It is heavier, higher-geared, and loses some power to the air-conditioning plant under its bonnet.

But don't be deceived: it can still knock spots off almost anything else on the road in various aspects of performance. It can go from rest to 80 in 6,5 seconds, and to 100 (still in 2nd) in 9,3. It covers a kilometre from rest in 29,6 seconds, and has topped a true 175 by the time it crosses the line. Running in its direct-drive fourth, it registered a true 206,3 km/h on a level road, both ways, pulling more than its rated 5 200 revs.

Understandably, there is not much pulling power in 5th (its gradient ability is 1-in-15,2) but this is a usable driving gear from about 60 km/h upwards — and is backed by tremendous climbing and overtaking ability in both 3rd and 4th.

The speedometer overreads by a reducing percentage as speed rises (4 per cent at 100, on the test car) and the rev-counter is extremely accurate right through the speed range.

ECONOMY AND SOUND

We have based our fuel consumption tables and graphs on the official ECE (Economic Community of Europe) figures, and these show that the car has the potential for amazing fuel economy at steady speeds. It will achieve 7,16 litres/100 km at 80 and 8,20 at 90 — setting new standards for any BMW model in current production, and giving it an 800 km cruising range.

But a word of caution: while our Fuel Index comes out at 9,31 litres/100 km, it is highly unlikely that anyone will drive this high-performance car in sedate style, and the ECE "urban cycle" figure goes as high as 18,5 litres/100 km.

The 3,5-litre fuel injected motor produces 160 kW and is teamed with an overdrive-style 5-speed gearbox.

Somewhere between those two — in the region of 14,0 litres/100 km — is probably what owners can expect, with the bonus of special economy on long-distance trips.

Mechanical sound levels are very modest for a performance car, thanks both to good build and long-legged gearing. Road rumble is higher than we expected, but the average of 79,2 decibels at 100 is satisfactory.

HANDLING AND BRAKING

The car's ride is not over-firm, but it is well-damped and there is no sloppy body movement in cornering and braking. We have always admired "5 Series" handling — BMW's best — and this new "5" flagship is a delight to drive and handle. Unlike most sports cars, it also has a commanding ride over dirt roads, and it takes loads with the best of family sedans.

Its brakes are high-performance, and the test car was even a shade over-braked at front. But it stopped firmly from high speeds, and braking actually improved as the discs warmed up.

TEST SUMMARY

The new M535i is difficult to categorise. It's not a pure sports sedan like the 530, nor is it pure luxury car. To some extent, it combines the best elements of both, to produce a really fast car with a high standard of comfort and equipment.

But whatever one's feelings on these aspects, there's no denying that it is a magnificent car: a blend of vigour and glamour, and very different from the general run of cars in its price class.

SPECIFICATIONS

ENGINE:

Cylinders	6 in line
Fuel supply	Bosch L-Jetronic fuel injection
Bore/stroke	93,4/84,0 mm
Cubic capacity	3 453 cm^3
Compression ratio	9,3 to 1
Valve gear	o-h-v, single o-h-c
Ignition	electronic
Main bearings	seven
Fuel requirement	98-octane Coast, 93-octane Reef
Cooling	water, thermo-coupled fan

ENGINE OUTPUT:

Max. power I.S.O. (kW)	160
Power peak (r/min)	5 200
Max. usable r/min	6 200
Max. torque (N.m)	310
Torque peak (r/min)	4 000

TRANSMISSION:

Forward speeds	five (four and overdrive)
Gearshift	console
Low gear	3,822 to 1
2nd gear	2,202 to 1
3rd gear	1,398 to 1
4th gear	direct
Top gear	0,813 to 1
Reverse gear	3,705 to 1
Final drive	3,07 to 1, ZF limited slip
Drive wheels	rear

WHEELS AND TYRES:

Road wheels	Mahle light alloy
Rim width	7,0J
Tyres	195/70 VR 14 steel radials
Tyre pressures (front)	220 to 260 kPa
Tyre pressures (rear)	250 to 290 kPa

BRAKES:

Front	280 mm discs, ventilated
Rear	272 mm discs
Pressure regulation	dual systems, anti-lock at rear
Boosting	vacuum servo
Handbrake position	between front seats

STEERING:

Type	ZF variable-rate power-assisted
Lock to lock	4,6 turns
Turning circle	11,1 metres

MEASUREMENTS:

Length overall	4,620 m
Width overall	1,690 m
Height overall	1,425 m
Wheelbase	2,636 m
Front track	1,460 m
Rear track	1,460 m
Ground clearance	0,140 m
Licensing mass	1 400 kg

SUSPENSION:

Front	independent
Type	coil struts, stabiliser bar, gas-filled shockabsorbers
Rear	independent
Type	coils, semi-trailing arms, stabiliser bar, gas-filled shockabsorbers

CAPACITIES:

Seating	4/5
Fuel tank	70 litres
Luggage trunk	430 dm^3 net

WARRANTY:

12 months.

TEST CAR FROM:

BMW South Africa, Rosslyn, Pretoria.

ACCELERATION

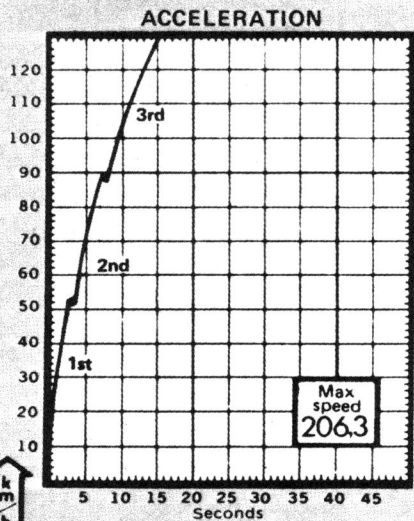

Max speed 206,3

BRAKING DISTANCES

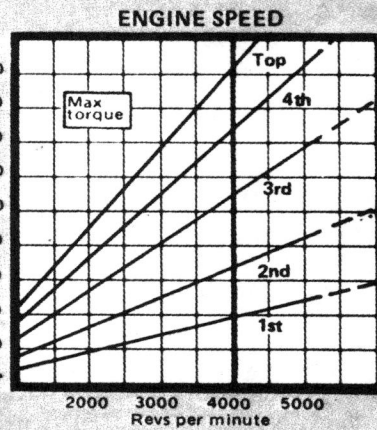

metres 30 40 50 60 70 80

1,0g   0,5g
(10 stops from 100 km/h)

NOISE VALUES

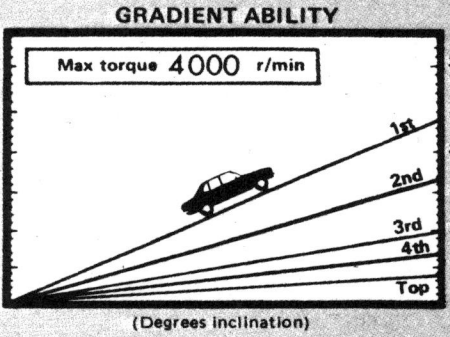

S.I.L.

dBA

MECH. WIND ROAD AVE.
(at 100 km/h)
S.I.L. = Speech interference level

CALCULATED FUEL RANGE
(km)

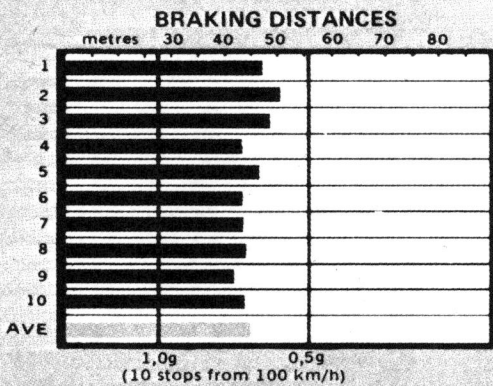

Tank 70ℓ

km

60 70 80 90 100
km/h

---

test BMW M 535i five-speed

PERFORMANCE

MAKE AND MODEL:
Make . . . . BMW
Model . . . . M 535i five-speed
PERFORMANCE FACTORS:
Power/mass (W/kg) net 114,3
Frontal area (m²) 2,41
km/h per 1 000 r/min (top) . . 46,4
INTERIOR NOISE LEVELS:

	Mech	Wind	Road
Idling	66,0	—	—
60	71,0	—	—
80	73,5	76,0	78,0
100	76,0	80,0	81,5
Average dBA at 100			79,2

ACCELERATION (seconds):
0-60 4,5
0-80 6,5
0-100 9,3
1 km sprint 29,6
OVERTAKING ACCELERATION:

	3rd	4th	Top
40-60	3,6	5,2	7,6
60-80	3,5	5,6	7,9
80-100 . . .	3,3	5,6	8,8

MAXIMUM SPEED (km/h):
True speed 206,3
Tachometer reading 5 000
Speedometer reading 216
Calibration:
Indicated: 60 70 80 90 100
True speed: 55 65 75,5 86 96
FUEL CONSUMPTION (litres/100 km):
60 6,51
70 6,79
80 7,16
90 7,60
100 8,20
BRAKING TEST:
From 100 km/h
Best stop 3,4
Worst stop 3,9
Average 3,59
GRADIENTS IN GEARS:
Low gear . . . . 1 in 2,2
2nd gear . . . . 1 in 3,4
3rd gear . . . . 1 in 5,8
4th gear . . . . 1 in 9,3
Top gear . . . . 1 in 15,2
GEARED SPEEDS (km/h):
Low gear 51,4
2nd gear 89,1
3rd gear 140,4
4th gear 196,3
Top gear 241,5
TEST CONDITIONS:
Altitude . . . . at sea level
Weather . . . . fine and hot
Fuel used . . . . 98 octane
Test car's odometer 18 010

ENGINE SPEED

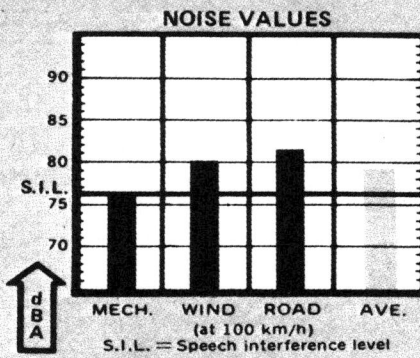

Max torque

Top 4th 3rd 2nd 1st

km/h

2000 3000 4000 5000
Revs per minute

IMPERIAL DATA

ACCELERATION (seconds):
0-60 8,7
MAXIMUM SPEED (m-p-h):
True speed 128,2
¡FUEL ECONOMY (m-p-g):
50 m-p-h 39,3
60 m-p-h 35,4
(¡Based on ECE figures.)

GRADIENT ABILITY

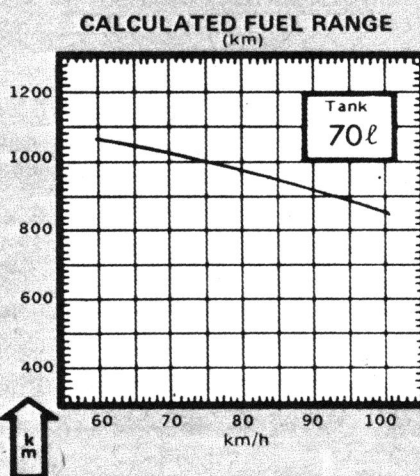

Max torque 4000 r/min

30° 25° 20° 15° 10° 5°

1st 2nd 3rd 4th Top

(Degrees inclination)

CRUISING AT 100

Mech noise level . . . . 76,0 dBA
0-100 through gears . . . . . 9,3 seconds
¡Litres/100 km at 100 8,20
Optimum fuel range at 100 . . 854 km
Braking from 100 . . . . . 3,59 seconds
Maximum gradient (top) . . 1 in 15,2
Speedometer error . . . . . . 4% over
Speedo at true 100 104
Tachometer error . . . . . negligible
Engine r/min at 100 . . . . . 2 155
(¡Based on ECE figures.)

STEADY-SPEED FUEL CONSUMPTION
(litres/100 km at true speeds)

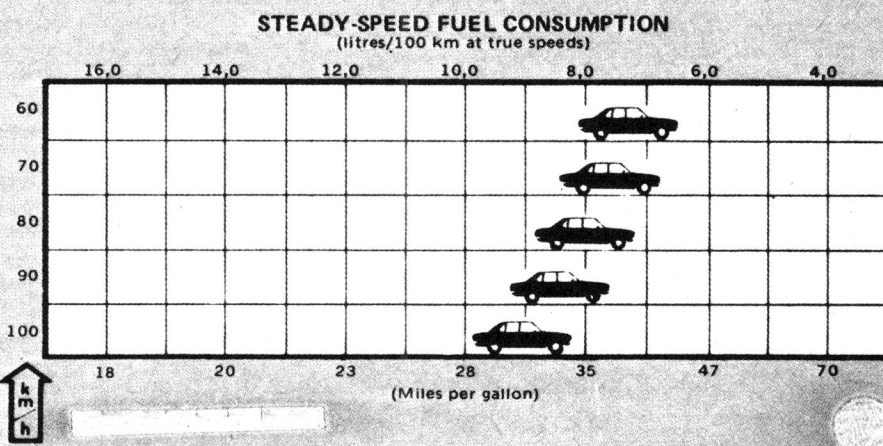

16,0 14,0 12,0 10,0 8,0 6,0 4,0

km/h

18 20 23 28 35 47 70
(Miles per gallon)

Understeer is rarely present in the M535i...

BMW M535i

The best yet from Bavaria?

THE BMW M535i follows a classic recipe for performance. Take the biggest engine in production then fit it into the smallest body shell feasible. The idea to produce an utterly refined up market car came from now departed competitions director Jochen Neerpasch who was in no doubt as to the performance potential of a car using mechanical components belonging to the 140 mph 635Csi but in a bodyshell weighing some 150lb less. The prefix M stands for Motorsport division where final setting-up and testing take place before delivery.

The engine is identical to that used in the 735i and 635Csi (not to be confused with the smaller bore 3.2-litre unit). In dry sump form the same siamesed block is also employed for the 24 valve 3.5-litre engine that powers the M1. Bore and stroke measurements of the single ohc unit are 93.4x84mm to give a displacement of 3,453 c.c. On a 9.3 to 1 compression ratio and L-Jetronic fuel injection maximum power and torque are 218 bhp (DIN) at 5,200, and 224 lb. ft. at 4,000 rpm.

Whereas normal production 5 series cars are available with four- or five-speed "overdrive" Getrag transmissions BMW have opted on the M535i to offer only the "sports" five-speed with direct top as fitted to the 635Csi. On a 3.07 final drive (again identical to the 635Csi) overall gearing is 23.5 mph per 1,000 rpm. Other parts common to the 635Csi are its 14 x 6½in. wide BBS/Mahle cast alloy wheels, 195/70VR Michilin XWX tyres, and all round vented discs.

Outwardly the main distinguishing features are its deep glass-fibre moulded front spoiler (vented to provide airflow to the brakes and oil cooler) and rear moulded rubber boot lid spoiler, both vital to achieve the necessary high-speed aerodynamic stability and balance. Underneath the skin suspension modifications include stiffer 528i "sportwagen" springs, harder damping (Bilstein) also stronger anti-roll bars front and rear measuring 24 and 18mm dia respectively. Inside there is a pair of high backed semi rally type seats and a smaller than standard flat Motorsport steering wheel.

When comparing weights the 535i's *raison d'etre* is confirmed. It tips the scales at 29.5cwt (distributed 55/44 front to rear) compared with our *Autotest* 635i's 30.8cwt. The saving is at least 140lb.

BMW M535i

UK appearance of four-cylinder 1,990 c.c. 520 in February 1973. December '73 saw introduction of carburettor 2,494 c.c. six cylinder 525, January '75 the 1,766 c.c. 518, April '75 the fuel injected 2,788 c.c. 528i and October '77 the 1,990 c.c. six-cylinder 520. The Motorsport conceived and "fettled" M535i was launched in Bavaria in March this year. In suspension, brakes and mechanics it has much in common with the 635Csi Coupé.

PRODUCED BY:
Bayerische Motoren Werke AG
Munich
West Germany

SOLD IN UK BY:
BMW (GB) Ltd
Ellesfield Avenue
Bracknell
Berks, RG12 4TA

Performance
Indecently fast

Starting is typical of L-Jetronic engines. With no throttle applied the engine fires almost immediately the key is turned and then settles down to an even tickover with that familiar BMW twitter from the exhaust. Warm-up is commendably fast. There is not the slightest suggestion of flat spotting or sluggishness, while the engine is reaching working temperature.

It is the smooth and totally fuss-free manner so typical of a BMW straight six that never fails to please. The car will trickle through traffic in top, then with under 1,000 rpm indicated pull away without a snatch or shudder. With 2,000 rpm showing it is getting into its stride hauling the car more rapidly forward, progress that seems not to diminish until well over 120 mph. The engine revs eagerly past peak power and on to the rev limiter (set at 6,200 rpm on the test car). Its beautifully progressive torque and power

mance of this measure makes it a most effortless car to drive fast. Acceleration response is superbly acute. Once the engine is in the working range — say above 2,500 rpm — down-changes are rarely needed to give adequate overtaking performance.

At 6,200 rpm, maxima in the gears are 39, 61, 82 and 116 mph — speeds that will vary slightly due to rev limiter tolerance. Using the car to the full one occasionally feels the need for a higher third. An 1,800 rpm rev drop occurs on the upchange to fourth yet there is only a 1,600 rpm gap between second and third. Ideally the reverse would be better, to preserve the otherwise progressive reduction in rev drops towards top. On the ''sports'' gearbox first gear is offset (whereas the ''overdrive'' boxes have a more conventional offset fifth). The first to second change is slow (making the 0-60 mph time even more remarkable) and a little indistinct. Once working in the normal H-pattern, change quality is accurate enough, but

should be an easily attainable average in normal use. The tank holds 15.5 gallons. On the test car the fuel warning light started flashing after covering between 210 and 230 miles leaving a comfortable four gallon reserve — and time to find a convenient filling station. Brimming is fast. The filler accepts full pump flow until attempts are made to squeeze in the last half gallon which has to be added slowly if blow backs are to be avoided. A nice point (on all BMWs) is a filler cap carrier mounted on the inside of the filler flap.

Noise
Mainly most pleasing ones

As mentioned the M535i will burble along with the town traffic, barely revving above tickover. On full throttle an induction chatter (well muted) signifies that the engine is beginning to get into its stride, and as the revs rise this blends into a delightful song. On cruising power settings mechanical noise

Road Behaviour
Secure ride

The M535i's power-assisted ZF recirculating ball steering is pleasing not only for its total absence of kickback but for its sensible gearing (3.5 turns), near perfect weighting and good feel (though this does not match the best rack and pinion systems). The car also has a reasonably tight 34ft turning circle.

Clearly BMW have chosen suspension settings for the M535i that equate closely to those for the 635CSi. It has much of that car's feel and virtually none of the more prosaic 5 Series model's tendency to pitch, and corkscrew when cornered or braked hard. This very fact leads to a secure yet comfortable ride. The suspension is firm yet supple enough to cushion the occupants from most sharp surface changes, though some vertical shaking was occasionally encountered over badly rippled B roads. Longer undulations and more

gressive torque and power curves are illustrated by the evenness of the 20 mph acceleration increments in any gear. In top they are within a second of each other from 20 mph right through to 110 — and from 10 mph to 100 mph in fourth. And as predicted by BMW the M535i *is* appreciably faster than the 635CSi (admittedly figured in less favourable conditions) everywhere and only marginally slower in top speed. 30-50 mph in third, 50-70 mph in fourth and 70-90 mph in fifth took 3.8, 6.0 and 8.7sec against 635CSi's 4.3, 6.7 and 10.0sec. The M535i took 13.8sec to accelerate from 100-120 mph in top carving some 4.2sec off the time achieved by our road test 635CSi (*Autocar*, 6 Jan, 1979).

Standing start acceleration is equally impressive. The limited slip differential gives superb traction off the line. Dropping the clutch with 4,500 rpm produced just the right amount of wheelspin, two black lines on the tarmac and 0-30 mph in 2.5sec. Thereafter the M535i stormed to 60, 100 and 120 in 7.1, 19.2 and 32.3sec respectively (635CSi 8.5, 23.4 and 41.0sec). For a ''family'' car the M535i is indecently fast. Perfor-

rather rubbery.

The M535i displays a degree of undergearing typical of many German cars where the ultimate in speed (and economy) have been sacrificed for top gear performance. We achieved a mean maximum of 139 mph (1 mph less than the 635CSi and well up to BMW's claimed 136 mph plus) with the engine revving at 5,900 rpm — some 600 rpm over peak power. Yet because of its turbine like smoothness the engine never feels stressed. Our testing was conducted in Germany where road conditions permitted quite natural cruising between 100 and 120 mph.

Economy
Not what you'd expect

Engine efficiency has always been a BMW strong point. In comparing the M535i's 20.2 mpg overall with our 635CSi 17.5 mpg, the suggestion is that light weight may have a greater bearing on economy than aerodynamic efficiency. For most of the test mileage the M535i was driven hard and rarely cruised below 90mph. On one long cross-country trip we saw nearly 22 mpg which

dies away except at around 4,000 rpm (95 mph) where a slight engine/body resonance makes its presence heard rather than felt. When cruising above 100 mph engine noise fades once more to be replaced by steadily increasing wind noise from around the screen areas as the most evident sound. Normal conversation is quite possible up to 100 mph, but needs slightly raised voices over 120 mph.

The M535i is particularly well-insulated from road noise. With VR rated tyres fitted some low speed bump thump over damaged road surfaces or potholes is to be expected, but it is always muted. At high speed coarse surfaces create some tyre roar but not to an intrusive degree. Paradoxically, we noticed that a marked howl eminated from the tyres at precisely 40 mph. At 45 mph it disappeared. While by no means quiet by Jaguar standards, it is the M535i's complete absence of engine or driveline vibration that adds to a feeling of overall refinement.

typical B road surface irregularities are absorbed beautifully. Pitch and roll are controlled sufficiently to inspire confidence without being so limited as to take away the driver's feel of the build up in braking and cornering forces.

On well surfaced roads straight line stability is excellent. Quite strong cross winds barely affect the car until travelling at over 120 m.p.h. However, driven fast on bumpy and cambered straights where suspension

movement is large the car would "walk" about slightly and thus require constant small steering corrections to keep it running straight. It should be emphasised that the car does not feel unstable in these conditions but its accurate placing requires concentration.

Motorsport appear to have struck just the right balance in matching the 5-series chassis to a large power increase. On dry roads the adhesion limit is excellent. The car responds keenly to the steering. There is no tugging the M535i into bends. Understeer is never present except perhaps on the entry to a hairpin bend, yet we have alway felt all BMWs could do with some to give the driver more idea of the degree of latent adhesion available — especially in the wet.

Pressed hard through long fast corners the front and rear tyres commence squealing together as if to emphasise the car's neutral balance. In easing off the throttle when hurrying through bends one feels some uneasiness from the rear, and an abrupt lift-off while cornering in extremis causes the quite sharp rear end breakaway so characteristic of BMWs — especially in the wet. This is in spite of the limited slip differential which offers such tenacious rear end bite out of wet surfaced corners. The trade off here being that when too much throttle is used rear end breakaway is not cus-

The engine bay is crammed full, yet all essentials are within easy reach. The six ram piped induction system presents a most purposeful sight. Nearest the camera are the water header tank, fuse container, and battery (deep down behind N/S headlight)

hioned (as it is normally) by a spinning inside rear wheel.

The brakes worked admirably during the fade test, coping easily with 10 consecutive 0.5g stops from 90 mph (the speed at the ¼-mile). After an initial rise pedal pressures stabilized at a consistently moderate level. Where the 535i's brakes do come in for criticism is in their excessively sharp response. A 10lb pedal load produced 0.25g retardation — the typical check

stop — while a mere 30lb (or just half what most cars require for a crash stop) saw our decelerometer record the best crash stop of 0.95g. With a 40lb pressure the front brakes locked hard on. Quite apart from the probability of locking-up during panic braking, such light pedal efforts are likely to lead to annoyingly uneven braking when using the brake pedal as a pivot during heel and toe brake and throttle operation. The hand-

brake coped well with a 1-in-3 test slope, and managed a tolerably good 0.3g stop on the flat.

Behind the wheel
Utterly straightforward

Inside, the M535i's main distinguishing feature is its superb high backed cloth covered and side bolstered semi-rally type seats: They have a huge range of fore and aft adjustment, and a ratchet type backrest movement. In addition the flat and slightly smaller than standard leatherbound Motorsport steering wheel is adjustable for reach. There should be little reason for discomfort.

The pedals are perfectly arranged for heel and toe operation. The instruments are housed in a hooded binnacle, and are plainly visible through the steering wheel. As we have consistently pointed out they are a model of simplicity and clarity. A 150 mph speedometer (with press to reset mileage trip) and 0-8,000 rpm revcounter (redlined at 6,400) are flanked by water temperature and fuel contents gauges. Disappointingly in a car of this class there is no oil pressure gauge or ammeter. Warning lights deal with these functions plus rear fog lights, handbrake, indicators and main beam. The push/pull main light and press button rear fog light switches are placed conveniently on the facia to the left of the steering column, with the hazard warning light switch opposite.

Thumb presses set in the steering wheel spokes operate the horn. We found them rather too close to the rim, and lost count of the times they were pressed inadvertently. Minor controls are crisp in use. The left hand stalk operates indicators, headlamp flash, dip and main beam, while the right hand one

Instruments from left are fuel gauge, speedometer (with press to re-set trip), revcounter, and water temperature gauge. Behind the flat Motorsport steering wheel (note horn thumb presses) column stalks operate indicators/headlamp flash on left, and wipers on the right. Centre console houses heater controls, cigar lighter, ashtray, and radio. The main light switch (push pull type) and hazard warning knob are placed beneath instrument panel on facia. Pedal positioning is excellent

works two speed plus intermittent action wipers, and presses at the end for screen wash and flick wipe.

Existing BMW owners will be entirely familiar with the heater controls mounted in the centre console. The outer dials control temperature and flow direction. Between these is the usual ring type fan rheostat (now with digital instead of analogue clock inset). Two horizontal sliders dictate whether the system draws in outside air or recirculates that within. Set in the fresh air mode, the system provides a good blast through four facia outlets though not without fan assistance to maintain flow at low speeds. Directional control and heat output are satisfactory, however once again we have to mention the poor temperature control offered by BMW's water-valve heater. This leads to constant fiddling in trying to "chase" the desired heat output.

One of the strongest features of the 5 series and the M535i in particular is its commanding driving position. All round visibility is first class, the squared off bonnet line making it a particularly easy car to judge widths in.

Living with the M535i

Owning a practical family car with more performance than is available in many a grand tourer will be rewarding experience, especially so as apart from the nicely finished aerodynamic add-ons, there is little to distinguish this from any other up market 5-series BMW. The driving seat must come in for a special mention. It gave tired testers complete support during long periods at the wheel.

Oddments space is tolerably good, with a lipped facia top,

generous door bins, a small tray in the centre console, and the usual drop down glovebox on the passenger's side which contains, at one side, a little torch that is automatically recharged from the car's electrical system. Other pleasing details are the automatic boot light, and comprehensive tool kit contained within a drop down tray in the boot lid. The jack, warning triangle, and a first aid kit are clipped to the boot side. The spare wheel is housed under the carpeted boot floor.

The standard specification includes tinted glass, electrically operated door mirrors and central door-locking, the latter only working from the driver's door key or catch. The test car was also fitted with a most satisfactory and worthwhile Webasto sliding and tipping sunshine roof.

Crammed as it is with engine and ancillaries, the M535i's engine bay presents a neat and pleasingly purposeful sight. Owners are unlikely to delve any deeper than undertake normal fluid level checks which are straightforward. Otherwise servicing follows normal BMW practice; there is an oil change every 5,000 miles and more major attention required at 10,000-mile intervals.

The 5 series range

Motorsport's intention is to produce 1,200 M535i models yearly. GB sales start in October and the price will be £13,745. Normal production models remain the 528i (£10,595), 525 (9,325), 520 (£8,150) and 518 (£6,958). Five-speed over-drive gearboxes are available on all models (£577 extra on 528i and 525, £515 and £274 more on the 520 and 518 respectively). "Sports" five-speed gearboxes can also be had for an additional £230 on all but the 518.

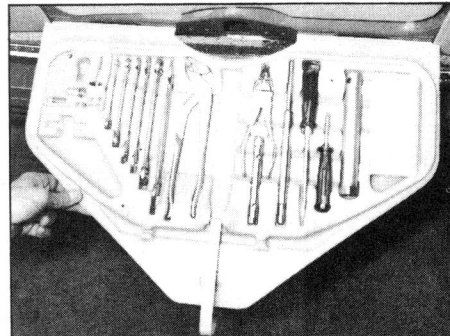

HOW THE BMW M535i PERFORMS

Figures taken at 6,242 miles by our own staff on the Continent

All Autocar test results are subject to world copyright and may not be reproduced in whole or part without the Editor's written permission

TEST CONDITIONS:
Wind: 0-5 mph
Temperature: 20 deg C (68 deg F)
Barometer: 29.5 in. Hg (1,000 mbar)
Humidity: 70 per cent
Surface: dry asphalt and concrete
Test distance: .859 miles

ACCELERATION

FROM REST

True mph	Time (sec)	Speedo mph
30	2.4	29
40	3.5	39
50	5.3	50
60	7.1	60
70	9.2	70
80	11.9	81
90	15.7	89
100	19.2	100
110	24.5	111
120	32.2	120
130	44.3	131

Standing ¼-mile: 15.7 sec, 90 mph
Standing km: 28.4 sec, 115 mph

IN EACH GEAR

mph	Top	4th	3rd	2nd
10-30	—	6.8	4.6	3.0
20-40	8.1	6.6	4.2	2.9
30-50	8.3	6.1	3.8	2.8
40-60	8.1	6.1	3.7	3.1
50-70	8.0	6.0	3.9	—
60-80	8.4	6.0	4.6	—
70-90	8.7	6.3	—	—
80-100	9.0	6.7	—	—
90-110	8.9	8.3	—	—
100-120	13.8	—	—	—
110-130	17.3	—	—	—

MAXIMUM SPEEDS

Gear	mph	kph	rpm
Top (mean)	139	224	5,900
(best)	140	225	5,950
4th	116	187	6,200
3rd	82	132	6,200
2nd	61	98	6,200
1st	39	63	6,200

FUEL CONSUMPTION

Overall mpg:
20.2 (14.1 litres / 100km)

Constant speed:
Autocar constant speed fuel measuring equipment incompatible with Bosch L Jetronic fuel injection

Autocar formula: Hard 18.2 mpg
Driving Average 22.2 mpg
and conditions Gentle 26.3 mpg

Grade of fuel: Premium, 4-star (98 RM)
Fuel tank: 15.5 Imp. galls (70 litres)
Mileage recorder: 1.7 per cent long

(ECE A70 fuel consumption figures: not necessarily related to Autocar figures)

Urban cycle: 14.7 mpg
Steady 56 mph: 33.1 mpg
Steady 75 mph: 26.8 mpg

OIL CONSUMPTION

(SAE 20/50) Negligible

BRAKING

Fade (from 90 mph in neutral)
Pedal load for 0.5g stops in lb

	start/end		start/end
1	20/16	6	24/48
2	20/40	7	28/42
3	22/60	8	28/44
4	22/100	9	36/40
5	24/60	10	32/36

Response (from 30 mph in neutral)

Load	g	Distance
10 lb	0.25	120 ft
20 lb	0.55	55 ft
30 lb	0.95	32 ft
Handbrake	0.30	100 ft
Max. gradient: 1 in 3		

CLUTCH
Pedal 35 lb; Travel 5 in

WEIGHT
Kerb, 29.5 cwt/3,305 lb/1,501 kg
(Distribution F/R, 55/45)
Test, 33.5 cwt/3,750 lb/1,703 kg
Max. payload 970 lb/440 kg

DIMENSIONS

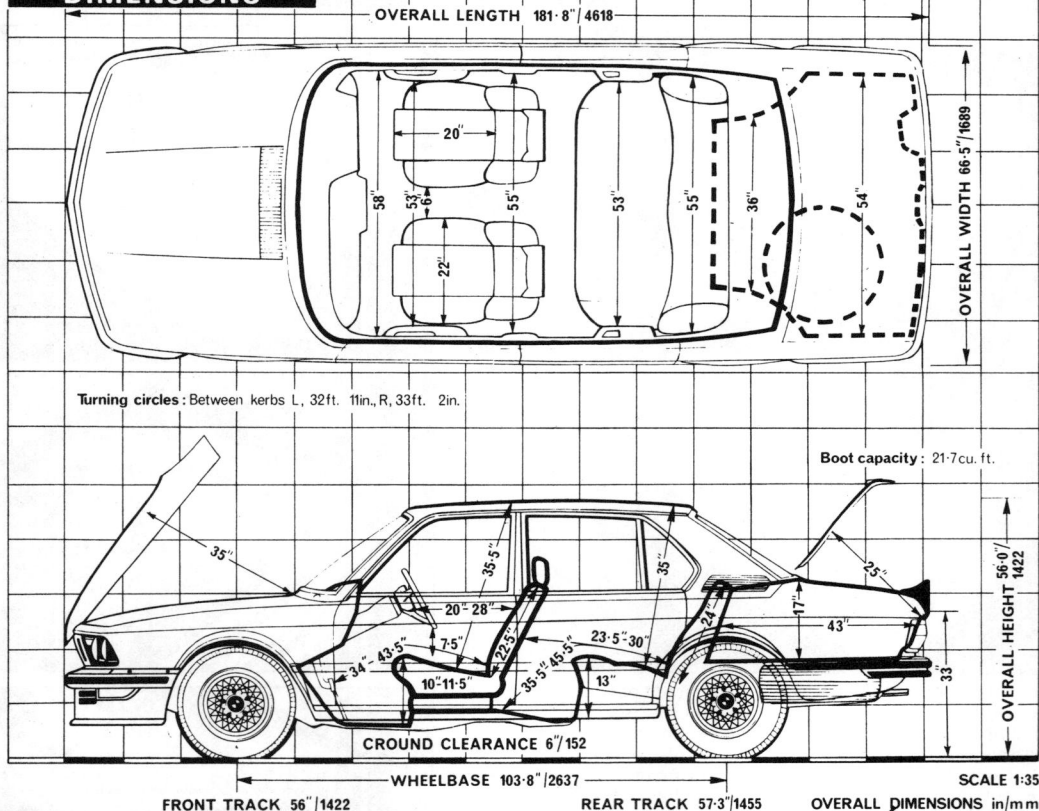

OVERALL LENGTH 181·8"/4618
OVERALL WIDTH 66·5"/1689
Turning circles: Between kerbs L, 32ft. 11in., R, 33ft. 2in.
Boot capacity: 21·7 cu. ft.
OVERALL HEIGHT 56·0"/1422
GROUND CLEARANCE 6"/152
WHEELBASE 103·8"/2637
SCALE 1:35
FRONT TRACK 56"/1422
REAR TRACK 57·3"/1455
OVERALL DIMENSIONS in/mm

PRICES

Basic	£11,032.7?
Special Car Tax	£919.4(
VAT	£1,792.8(
Total (in GB)	**£13,744.9?**
Seat Belts	Incl
Licence	£60.0(
Delivery charge (London)	£80.0(
Number plates	£10.0(
Total on the Road	**£13,894.9?**
(exc. insurance)	

EXTRAS (inc. VAT)
*Sunshine roof (manual)	£410.5?
*Sunshine roof (electric)	£566.9?
*Radio/cassette player	£120.00 (approx
Air conditioning	£1,130.0(
*Fitted to test car	

TOTAL AS TESTED ON THE ROAD £14,425.5?

Insurance   Group 7/On application

SERVICE & PARTS

Interval

Change	5,000	10,000	20,00?
Engine oil	Yes	Yes	Yes
Oil filter	Yes	Yes	Yes
Gearbox oil	—	—	Yes
Spark plugs	—	Yes	Yes
Air cleaner	—	Yes	Yes

Total cost
(Assuming labour at £11.50/hour inc. VAT)

PARTS COST (including VAT)
Brake pads (2 wheels)—front	£19.7(
Brake pads (2 wheels)—rear	£11.6?
Exhaust complete	£181.0?
Tyre — each (typical)	£102.9?
Windscreen (tinted and laminated)	£79.5?
Headlamp unit	£17.4?
Front wing	£84.0(
Rear bumper (3 sections)	£109.6?

WARRANTY
12 months/unlimited mileage

SPECIFICATION

ENGINE
Head/block	Alloy/cast iron
Cylinders	6
Main bearings	7
Cooling	Water
Fan	Viscous
Bore, mm (in.)	93.4 (3.68)
Stroke, mm (in.)	84.0 (3.31)
Capacity, cc (in³)	3,453 (210.7)
Valve gear	Ohc
Camshaft drive	Chain
Compression ratio	9.3-to-1
Ignition	Electronic breakerless
Fuel injection	Bosch L-Jetronic
Max power	218 bhp (DIN) at 5,200 rpm
Max torque	224 lb ft at 4,000 rpm

ENGINE Front, rear drive

TRANSMISSION
Type	Getrag five speed
Clutch	Hydraulic diaphragm spring

Gear	Ratio	mph/1000rpm
Top	1.0	23.55
4th	1.263	18.65
3rd	1.776	13.26
2nd	2.403	9.80
1st	3.717	6.34

Final drive gear   Hypoid bevel
Radio   3.07 to 1

SUSPENSION
Front—location	MacPherson strut
springs	Coil
dampers	Telescopic
anti-roll bar	Yes (24 mm)
Rear—location	Independent semi-trailing arm
springs	Coil
dampers	Telescopic
anti-roll bar	Yes (18 mm)

STEERING
Type	ZF recirculating ball
Power assistance	Yes
Wheel diameter	14 in.
Turns lock to lock	3.5

BRAKES
Circuits	Twin split front and rear
Front	11.0 in. dia. ventilated disc
Rear	10.7 in. dia. ventilated disc
Servo	Vacuum
Handbrake	Centre lever, rear drum within disc

WHEELS
Type	Alloy
Rim Width	6½ in.
Tyres—make	Michelin XWX
type	Radial ply
size	195/70 VR 14
pressures	F 30, R 30 psi (normal driving)

EQUIPMENT
Battery	12V 66Ah
Alternator	65A
Headlamps	110/220W
Reversing lamp	Standard
Hazard warning	Standard
Electric fuses	17
Screen wipers	2-speed + intermittment + flick wipe
Screen washer	Electric
Interior heater	Water valve
Air conditioning	Extra
Interior trim	Cloth seats, pvc headlining
Floor covering	Carpet
Jack	Screw pillar
Jacking points	2 each side under sills
Windscreen	Laminated
Underbody protection	Paint system, bitumastic, pvc

26

BMW M535i £13,745

Front engine,
rear drive

Capacity
3,453 c.c.

Power
218 bhp (DIN)
at 5,200 rpm

Weight
2,971 lb / 1,350 kg

Autotest
30 August 1980

Audi 200 5T £12,950

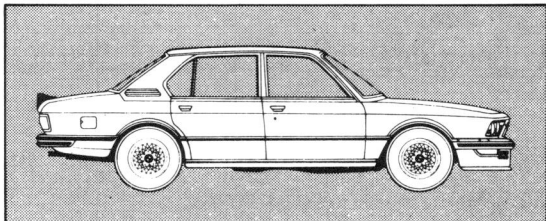

Front engine,
front drive

Capacity
2,144 c.c.

Power
170 bhp (DIN)
5,300 rpm

Weight
2,910 lb / 1,320 kg

Autotest of Audi 200
5T Automatic
July 1980

Ford Granada 2.8i GLS £9,300

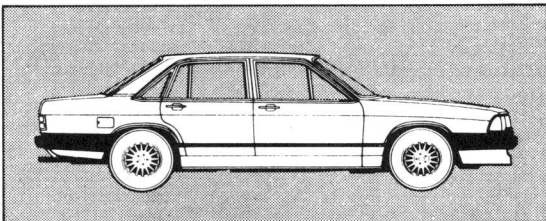

Front engine,
rear drive

Capacity
2,792 c.c.

Power
160 bhp (DIN)
at 5,200 rpm

Weight
2,971 lb / 1,350 kg

Autotest of 2.8iS
18 February 1978

Mercedes-Benz 280SE £14,458

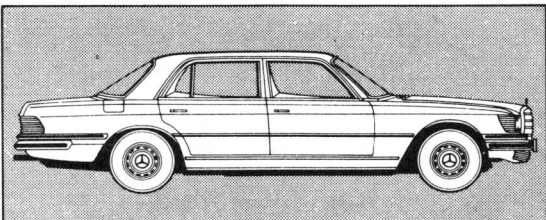

Front engine,
rear drive

Capacity
2,748 c.c.

Power
185 bhp (DIN)
at 6,000 rpm

Weight
3,676 lb / 1,669 kg

Autotest
12 July 1973

Opel Senator 2.8S £9,223

Front engine,
rear drive

Capacity
2,784 c.c.

Power
180 bhp (DIN)
at 5,800 rpm

Weight
3,080 lb / 1,400 kg

Autotest of Senator
automatic
1 November 1978

Saab 900 Turbo £10,750

Front engine,
front drive

Capacity
1,985 c.c.

Power
145 bhp (DIN)
at 5,000 rpm

Weight
2,888 lb / 1,310 kg

Autotest
28 July 1979

MPH & MPG

Maximum speed (mph)

BMW M535i	139
Audi 200 5T*	123
Mercedes Benz 280 SE*	120
Opel Senator*	119
Saab 900 Turbo	118
Ford Granada 2.8i S	117

Acceleration 0-60 (sec)

BMW M535i	7.1
Audi 200 5T*	8.7
Ford Granada 2.8iS	8.9
Opel Senator*	9.2
Saab 900 Turbo	9.6
Mercedes Benz 280 SE*	9.7

Overall mpg

Saab 900 Turbo	22.2
Ford Granada 2.8iS	20.8
BMW M535i	20.2
Opel Senator*	18.6
Audi 200 5T*	17.7
Mercedes Benz 280 SE*	16.7

*Test data for automatic versions

Other possible choices might be the Rover 3500S (£11,852) and Jaguar XJ6 4.2 (£15,798). Neither would come near to beating the 535i. The old maxim is confirmed; put big engine in tolerably light medium sized car for vivid performance and relative economy. Even if manual gearboxed test data were available for the Opel and Audi Turbo (there is no manual gearbox option on the Mercedes), the BMW would still destroy the competition. It is considerably more accelerative in the mid range than the 635 Csi coupé (and only 1 mph slower ultimately). One would have to look to the barely faster and much thirstier Porsche 928 or Jaguar XJ-S Auto for a true mid-range performance comparison. Best of all the BMW's performance comes with very impressive levels of mechanical refinement only equalled by the Saab (another undeniably efficient car), Opel or Mercedes.

ON THE ROAD

All steer well and are straight-line stable. Assuming Opel and Saab buyers go for the optional Pirelli P6 package to match the P6 and TRX shod Audi and Granada it would be hard to split these four on ultimate *dry* road grip. Understeer is of course more pronounced on the front-wheel-drive pair and traction sometimes hard to find (particularly in the Audi) on wet roads. The limited slip equipped M535i rates highly on traction. Understeer is rarely present. Its eager turn-in and neutral-power-on cornering balance are typically BMW, i.e. in extremis lift off oversteer is present. Power oversteer is easily provokable in the lower gears on wet roads. The Granada, Opel, and Mercedes also have semi-trailing arm rear suspension but have much less power and are more docile to handle

at the limit having some initial understeer and mild lift-off ''tuck . in''. On ride the Opel and Mercedes score high with the tauter BMW, S-pack Granada, Audi, and Saab marginally less compliant but still comfortable. The BMW's brakes are powerful but oversensitive. A criticism that applies to a lesser extent in the Granada and Mercedes. All are well enough ventilated, and have adequate heat supply. In heater control, the BMW's water valve system comes firmly last.

SIZE & SPACE

Legroom front / rear (in)

(seats fully back)

Mercedes Benz 280 SE	42 / 40
Ford Granada 2.8i S	41 / 40
Audi 200 5T	42 / 37
BMW M535i	43.5 / 35.5
Opel Senator	39 / 39
Saab Turbo	38 / 39

All six cars provide comfortable and supportive seating, also roomy accommodation for four. We liked the BMW particularly for its quite superb semi-rally type seats (peculiar to the M535i) and their huge range of adjustment, though when moved back fully this and their bulk inevitably restricts rear seat legroom somewhat.
The Saab provides accommodation enough to belie its overall compactness — and the above figures — moreover it has the advantage of having hatchback load-carrying versatility. Otherwise there is little to choose in luggage carrying capacity with the proviso that vertically mounted spare wheels in the Opel and Granada slightly restrict usable boot width.

VERDICT

If performance be your main criteria then the M535i stands alone and unequalled. It evokes something of the old 3.0-litre CSi lightweight Coupe's character — its vivid acceleration and mechanical smoothness. The S-class Mercedes (available with automatic only) has a solid quality. It more than matches the BMW in high speed cruising refinement, stability, handling, but needs an engine of some 4-litres to come near the Bavarian's performance. For the less well heeled buyer (comparatively) who simply requires a car that is quick enough and in other respects an excellent all rounder, the Opel (or Vauxhall Royale) and Granada offer much. Equally practical though perhaps less appealing to the real enthusiast are the front wheel drive Saab and Audi Turbos. Frankly all are desirable. Yet for those who can afford the extra (yet nearly £3,000 less than the 635CSi) there is this remarkable full family sized BMW. It possesses real grand touring car performance, better than usual economy, refinement, and manners that, if not perfect, come quite close.

BMW M535i

The 3.5-litre engine has been squeezed into the 5-Series body to produce a home-grown Alpina competitor. Performance is fine, but handling is a disappointment

YOU COULD argue that BMW's thunder has been well and truly stolen. As good ideas go, the marriage of the Bavarian company's 5-Series bodyshell and muscular fuel-injected 3.5-litre straight-six is old hat. Such a car has existed in the Alpina-prepared B9 for a couple of years, and with around 12 per cent more power than the standard engine's 218 bhp — and Alpina's new B10 has 260 bhp.

It is doubtful whether BMW has lost much sleep over this. Alpina, after all, is just another customer. And if BMW's image has been enhanced by its activities then so much the better. But it would be naive to suppose that those activities haven't had a bearing on the way BMW's own long-awaited 3.5-litre 5-Series models — the be-spoilered M535i and the plainer 535i — have turned out. To have clashed head-on with Alpina would have been an obvious mistake, though it's clear that with the dressed-up "Motorsport" 535i BMW is out not only to sharpen 5-Series commercial appeal in the face of Mercedes' 200/280 model onslaught, but to grab a slice of Alpina's action. The big difference, of course, is price. While the £24,995 Alpina B10 claims to be the swiftest production four-door saloon available in Britain, the M535i we test here claims to be the swiftest at less than £20,000.

In fact, at £17,950 the M535i — for which BMW claims a top speed of 143 mph and 0-60 mph acceleration of 7.0 sec — misses the £20,000 barrier by a tidy margin. But in the *fastest-four-door* contest, it's up against formidable opposition, not all of it with road-racer aspirations. Audi's 200 Turbo (£19,357), Citroën's CX GTi Turbo (£12,990), Jaguar's V12 Sovereign HE (£21,995) and Rover's Vitesse (£15,465) are all capable of seating four adults in comfort at speeds of over 130 mph. Volvo's 760 Turbo (£14,850) isn't quite as quick, but still totes a healthy performance image in a large, luxury package.

None of these cars has the sporting chic of the M535i. For the fashion-conscious, there are the obligatory spoilers and side-skirts, though these play their part in keeping the drag coefficient down to a respectable 0.37. Distinctive 165 TR 390 M-Technics light alloy wheels wearing 220/55 VR 390 Michelin TRX tyres provide further physical demarcation between the 5-Series flagship and its lesser stablemates.

What casual observation can't reveal are the revised spring and damper rates, limited slip differential and ABS anti-lock braking system that comes as part of what BMW calls its "performance" package. Cabin standard equipment includes electric windows, central locking, twin heated door mirrors, a headlight wash-wipe system, power steer-ing and an on-board computer.

Such are the trappings of the M535i's power-with-prestige image. The power, of course, is provided by BMW's 3430 cc alloy-headed six-cylinder engine lifted, without modification, from the 635 CSi and 735i models. On a compression ratio of 10:1 and with digital electronics to manage both the ignition and fuel injection, the unit develops 218 bhp (DIN) at 5200 rpm with 224 lb ft of torque at 4000 rpm. The engine drives to the rear wheels via either a five-speed manual transmission (our test car was fitted with the optional close-ratio unit) or a four-speed automatic with switchable "sports" and "economy" modes. Otherwise, the mechanical specification is pure 528i with all-independent suspension by Mac-Pherson struts and double pivot linkage at the front, semi-trailing arms at the rear and anti-roll bars at both ends, power-assisted recirculating ball steering and disc brakes all round, ventilated at the front.

No one could call the 130 mph 528i a slow car; it's just that the M535i is a very much quicker one — as, indeed, you would expect it to be with an extra 34 bhp to offset a modest 1.4 cwt (70 kg) increase in weight. Round Millbrook's high-speed

bowl it clocked 141.6 mph, with a best down-wind leg of just over the 143 mph BMW claims for it. On the flat and in still air, 143 mph would be about right with the close-ratio equipped car pulling a little more than 6000 rpm in fifth (23.5 mph/1000 rpm). Of the selected rivals we've maximum-speed tested, only the 139.5 mph Audi 200 Turbo gets close to the M535i flat out.

Not so, however, in a standing start sprint. Here, the muscular BMW asserts its authority in no uncertain terms, laying rubber for fully 30 yards before first winds out to almost 40 mph. There's a solid thump in the back as the clutch bites in second and the M535i surges towards the 60 mph benchmark in 6.9 sec. The push is barely diminished as third red-lines at 6000 rpm and 80 mph, while fourth sees the

speedometer needle sweep past the ton, our electronic test gear freezing the clock at 18.3 sec. There's no question about it, the M535i is a rapid car through the gears, capable of seeing off any of our selected rivals. But the fourth and fifth gear flexibility the big engine endows is, if anything is even more impressive: in fourth, the 3.5-litre Bee Emm can cover all the 20 mph increments between 20 and 90 mph in under six seconds, the important 40-60 and 50-70 mph sprints being disposed of in a mere 5.2 and 5.3 sec apiece. In fifth, these increments take 7.3 and 7.2 sec respectively.

On the road, this is never less than apparent, the M535i achieving in fifth what the average hot hatchback would be hard-pushed to manage in third. Yet in purely subjective terms, the engine failed to measure up to BMW's own high standards. Perhaps the unit was unrepresentative, but it revved with neither the silkiness nor the eagerness we had been expecting. Above 5500 rpm, the engine simply sounded strained and, even at lower revs, it lacked that crisp sporting edge so characteristic of BMW sixes. Yet no one could accuse the engine of being

noisy at any point in its rev range.

With an overall consumption of 19.7 mpg, the M535i isn't going to win any prizes for efficiency against the remarkable Audi 200 Turbo (23.5 mpg), but judged by the 20 mpg norm for a class composed of slower rivals, it's an acceptable enough result. Given a modicum of restraint, most owners should experience little trouble in matching, or even bettering, our projected touring figure of 24.5 mpg which gives a maximum range in excess of 370 miles on a 15.4-gallon (70 litre) tankful of four-star.

The quality of the ZF shift leaves something to be desired. It wasn't the dog-leg first that our testers objected to so much as the rather sloppy and ill-defined across-gate movements of the stubby gearlever. During the acceleration runs at Millbrook, it was all too easy to shift straight from first to fourth. Nor did the clutch elicit any great enthusiasm, having a rather long-winded action yet a late and abrupt take-up.

On dry roads, the M535i handles neatly and securely with enough grip in reserve to satisfy the most enthusiastic drivers. The steering is particularly easy to come to terms with being quick (despite an overlarge "sports" steering wheel), well-weighted and properly communicative. The car turns crisply and hold a chosen line with fine accuracy, even if the road is bumpy. Mild understeer in tight turns changes to a more neutral balance with speed and although severe bumps and dips can introduce an element of diagonal pitching when pushing really hard — a case for firmer damping than Motorsport's engineers deemed necessary — the BMW's composure is never seriously upset.

Considering all of the above, it would be hard for even the most sensitive and chassis-conscious driver to predict the M535i's behaviour in wet or greasy conditions. And that's the problem. In stark contrast to the well-mannered and predictable wet-weather handling exhibited by the 528i — and even the hugely powerful Alpina B9 — the M535i is an unpleasant revelation, mimicking undesirably tail-happy traits that marred big-engined BMWs of old. A surprising degree of lift-off oversteer is part of the concern, here, but more worrying is the ease with which rear-end traction can be broken under power. Even the most ardent opposite-lock merchants would tire eventually of the level of concentration needed to make brisk progress in anything less than completely dry conditions, especially since the breakaway is by no means that progressive and requires quick reactions and accurate inputs to correct neatly.

The reasons for the M535i's dramatically different handling balance at first seem hard to pin down. But apparently small changes can have a disproportionately large effect. In the M535i's case, the root of the problem would seem to be two-fold, with each aspect contributing to increased oversteer. First, the Motorsport car has a rear anti-roll bar which is slightly *stiffer* (15.5 mm diameter against 13 mm) than the standard 528i's. In the interests of better traction in corners and more neutral handling balance, Alpina opted to go in the opposite direction, fitting a significantly softer 12 mm roll bar to the rear of the B9. Second, if we continue the comparison with the B9 — which, interestingly, uses identical rate gas-filled Bilstein

Above: Facia looks good, controls work well. Left: Clear, comprehensive instrumentation. Below: Body-hugging seats

dampers all round — we find that the M535i's ride height is lower, which means that the 13 deg trail angle of the trailing arm rear suspension effectively introduces more negative camber to the rear wheels, lessening the contact area of the wide tyres.

Alpina chooses Pirelli P7s; Michelin TRXs are worn by the M535i. Whatever the difference, the Motorsport BMW disappoints because it lacks the fine balance, poise and ultimate security of the excellent B9.

Some compensation for the wayward handling is offered by the ride which, although notably firm around town, falls some way short of being harsh. And at cruising speeds, the suspension copes admirably with a wide variety of surface conditions, soaking up even quite severe crests and ruts with well-damped aplomb. We were less enamoured with the brakes which, although equipped with the ABS anti-lock system as standard, were prone to judder after repeated hard use and, even in normal driving, were handicapped by a decidedly inert pedal response.

But for the superbly Recaro front seats, chunky sports steering wheel and rather overbearing use of black trim for the interior, the M535i otherwise appears very similar to its 528i counterpart. It's a design that encapsulates reasonable accommodation for a quartet of adults, a big boot, text-book ergonomcis and fine-all-round visibility.

In the BMW tradition, instrumentation scores for being both informative and beautifully presented, while the second generation on-board computer

fitted as standard is a lot easier to use than its predecessor, thanks to a remote push-push button on the end of the left-hand column stalk which allows the computer's functions to be reviewed in sequence. As we've mentioned many times before, 5-Series models are blessed with excellent heating and ventilation systems and, in terms of assembly and finish, build-quality is hard to fault.

The M535i is a car about which it is almost impossible not to have high hopes. And, as the performance figures show, the Motorsport car isn't short of basic pace. Indeed, on a performance-per-£ basis, it rates as fine value. Yet ultimately it is a disappointment, a combination of impressive credentials that haven't been tied together with sufficient thought or flair. The tricky wet-weather handling is a glaring flaw, the awkward gearchange a constant irritation. Improvements in both these key areas would make the M535i exactly what it should be: a cut-price Alpina. Until then, we'll stick with the 528i and keep saving.

Make: BMW **Model:** M535i
Country of Origin: Germany
Maker: Bayerische Motoren Werke AG, 8000 Munich 40, West Germany
UK concessionaire: BMW (GB) Ltd, Ellesfield Avenue, Bracknell, Berkshire RG12 4TA. Tel: 0344-26565
Total Price: £17,950.00
Options: Air conditioning £1401, Leather upholstery £795, Electric sunroof £611, Electric front seat adjustment £571, Metallic paint £338, Cruise control £264

MOTOR ROAD TEST
BMW M535i

PERFORMANCE

WEATHER CONDITIONS

Wind	0-5 mph
Temperature	8 deg F/46.4 deg C
Barometer	30.1 in Hg/1019 mbar
Surface	Dry tarmacadam

MAXIMUM SPEEDS

	mph	kph
Banked Circuit (5th gear)	141.6	227.8
Best ¼ mile (5th gear)	143.1	230.2
Terminal speeds:		
at ¼ mile	92	148
at kilometre	114	183
Speeds in gears (at 6000 rpm):		
1st	38	61
2nd	59	94
3rd	80	129
4th	112	180

ACCELERATION FROM REST

mph	sec	kph	sec
0-30	2.5	0-40	2.1
0-40	3.8	0-60	3.3
0-50	5.1	0-80	5.0
0-60	6.9	0-100	7.2
0-70	8.7	0-120	9.8
0-80	11.5	0-140	13.5
0-90	14.4	0-160	18.2
0-100	18.3	0-180	27.2
0-110	24.7		
Stand'g ¼	15.2	Stand'g km	28.0

ACCELERATION IN TOP

mph	sec	kph	sec
20-40	7.3	40-60	4.5
30-50	7.3	60-80	4.4
40-60	7.3	80-100	4.6
50-70	7.2	100-120	4.5
60-80	7.4	120-140	5.1
70-90	8.0	140-160	5.6
80-100	8.6	160-180	8.0
90-110	10.4		

ACCELERATION IN 4TH

mph	sec	kph	sec
20-40	5.7	40-60	3.6
30-50	5.4	60-80	3.3
40-60	5.2	80-100	3.3
50-70	5.3	100-120	3.4
60-80	5.4	120-140	3.6
70-90	5.9	140-160	4.9
80-100	7.3		
90-110	9.9		

FUEL CONSUMPTION

Overall	19.7 mpg
	14.3 litres/100 km
Touring*	24.5 mpg
	11.5 litres/100 km
Govt tests	17.7 mpg (urban)
	34.0 mpg (56 mph)
	28.5 mpg (75 mph)
Fuel grade	97 octane
	4 star rating
Tank capacity	70 litres
	15.4 galls
Max range*	377 miles
	607 km
Test distance	1013 miles
	1630 km

*Based on official fuel economy figures — 50 per cent of urban cycle, plus 25 per cent of each of 56/75 mph consumptions

STEERING

Turning circle	9.9 m, 32.8 ft
Lock to lock	3.1 turns

NOISE

	dBA
30 mph	63
50 mph	67
70 mph	72
Maximum†	77

†Peak noise level under full-throttle acceleration in 2nd

SPEEDOMETER (mph)

True mph	30	40	50	60	70	80	90	100
Speedo	31	41	52	62	72	82	92	102

Distance recorder: 1.2 per cent fast

WEIGHT

	kg	cwt
Unladen weight*	1391	27.4
Weight as tested	1579	31.1

*No fuel

Performance tests carried out by *Motor*'s staff at the Motor Industry Research Association proving ground, Lindley, and Millbrook proving ground, near Ampthill.

Test Data: World Copyright reserved. No reproduction in whole or in part without written permission.

GENERAL SPECIFICATION

ENGINE

Cylinders	6 in-line
Capacity	3430 cc
Bore/stroke	92/86 mm
Max power	218 bhp 163 KW at 5200 rpm (DIN)
Max torque	224 lb ft 304 Nm at 4000 rpm (DIN)
Block	Cast iron
Head	Aluminium alloy
Cooling	Water
Valve gear	Sohc, 2 valves per cylinder, chain drive
Compression	10.0:1
Fuel system	Bosch ME Motronic Fuel injection
Ignition	Programmed electronic
Bearings	7 main

TRANSMISSION

Drive	To rear wheels
Type	5-speed, manual
Internal ratios and mph/1000 rpm	
Top	1.00:1/23.5
4th	1.26:1/18.6
3rd	1.77:1/13.3
2nd	2.40:1/9.8
1st	3.72:1/6.3
Rev	4.23:1
Final drive	3.07

AERODYNAMICS

Coef. Cd	0.37

SUSPENSION

Front	Independent by MacPherson struts, coil springs, double pivot linkage, anti-roll bar
Rear	Independent by semi-trailing arms, coil springs, braking/lift-off compensator, anti-roll bar

STEERING

Type	Recirculating ball
Assistance	Yes

BRAKES

Front	Ventilated discs, 28.5 cm dia
Rear	Discs, 28.5 cm dia
Servo	Yes
Circuit	Diagonal split
Rear valve	Yes

WHEELS/TYRES

Type	Alloy, 165 TR 390 mm dia
Tyres	220/55 VR 390 TRX
Pressures F/R (normal)	33/36 psi 2.3/2.6 bar
(full load/ high speed)	36/42 psi 2.5/3.0 bar

ELECTRICAL

Battery	12V, 66 Ah
Alternator	80 Amp
Fuses	17
Headlights	
type	Halogen
dip	110 W total
main	230 W total

GUARANTEE

Duration	12 months unlimited mileage
Rust warranty	6 years against perforation corrosion

MAINTENANCE

Servicing	At intervals shown by service indicator

The Rivals

Other possible rivals include the Jaguar XJS 3.6 (£19,248), the Mercedes 380 SE (£22,355), and the Saab 900 Turbo 16 S (£14,090)

BMW M535i — £17,950

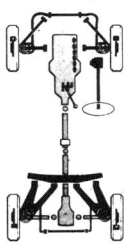

Price £	17,950
Capacity, cc	3430
Power, bhp/rpm	218/5200
Torque, lb ft/rpm	224/4000
Max speed, mph	141.6
0-60 mph, sec	6.9
30-50 mph in 4th, sec	5.4
mph/1000 rpm	23.5
Overall mpg	19.7
Touring mpg	24.5
Weight, kg	1391
Drag coefficient Cd	—
Boot capacity m³	0.37

Length 182" Width 67" Front track 56.3"
Wheelbase 103.3" Height 51" Rear track 57.5"

M535i has all the right ingredients to make it a cut-price Alpina, but slips up dynamically with wet weather handling that is painfully reminiscent of old big-engined BMWs. On a more positive note, straight-line performance is terrific (though engine lacks crispness) and economy very fair. Usual 5-Series strengths: room, build, comfort. And, at under £18,000, the M535i must rate as good value.

AUDI 200 TURBO — £19,357

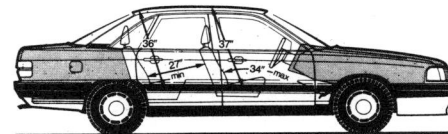

Price £	19,357
Capacity, cc	2144
Power, bhp/rpm	182/5700
Torque, lb ft/rpm	186/3600
Max speed, mph	139.5
0-60 mph, sec	7.5
30-50 mph in 4th, sec	7.2
mph/1000 rpm	23.5
Overall mpg	22.5
Touring mpg	29.3
Weight, kg	1362
Drag coefficient Cd	0.33
Boot capacity m³	0.49

Length 189" Width 71" Front track 57.8"
Wheelbase 105.8" Height 56" Rear track 57.8"

Based on the 100, the blown 200 achieves superb performance and a very impressive top speed from only 2.1 litres. This is combined with excellent economy. Precise handling, strong grip and outstanding stability are further plus points, as are the generous accommodation and fine finish. Ride and refinement are slightly disappointing. Standard equipment includes ABS, air conditioning, central locking and electric windows. Better in automatic form.

CITROËN CX GTi TURBO — £12,990

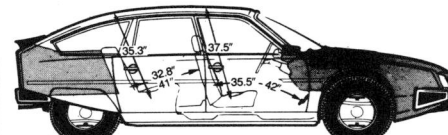

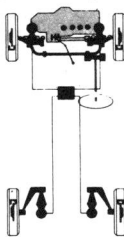

Price £	12,990
Capacity, cc	2473
Power, bhp/rpm	168/5000
Torque, lb ft/rpm	217/3250
Max speed, mph	129.6
0-60 mph, sec	7.6
30-50 mph in 4th, sec	5.1
mph/1000 rpm	25.2
Overall mpg	20.0
Touring mpg	26.0
Weight, kg	1390
Drag coefficient Cd	0.36
Boot capacity m³	0.29

Length 181" Width 68" Front track 58"
Wheelbase 112" Height 53.5" Rear track 53.5"

Not quite the fastest Citroën ever, the CX GTi Turbo is certainly the quickest-accelerating. Massive mid-range torque means that Citroën's new flagship is very good across the ground and a fine 120 mph cruiser, though economy is only average. Dynamic ability is well up to the performance but CX isn't the easiest of cars to drive smoothly. Suberb ride and comfortable interior are further plus points, but gearchange and boot are average. Very good value.

JAGUAR SOVEREIGN HE — £21,995

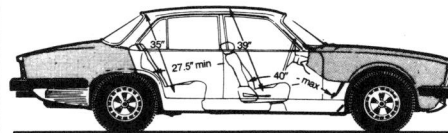

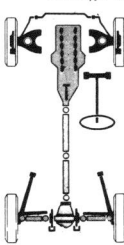

Price £	21,995
Capacity, cc	5345
Power, bhp/rpm	299/5500
Torque, lb ft/rpm	318/3000
Max speed, mph	145e
0-60 mph, sec	8.1
30-50 mph in kickdown, sec	3.1
mph/1000 rpm	26.8
Overall mpg	15.6
Touring mpg	19.6
Weight, kg	1915
Drag coefficient Cd	—
Boot capacity m³	0.33

Length 195.2" Width 69.7" Front track 58"
Wheelbase 112.8" Height 54" Rear track 58.5"

Modifications to the magnificent V12 engine improved fuel economy without compromising its exhilarating high-speed performance, but it's still a thirsty car. Available only in automatic form, the Jaguar offers unsurpassed levels of refinement, and an excellent ride combined with safe and predictable handling. Over-light power steering may not be to some drivers' tastes and the instrumentation looks dated. Superb finish and sumptuously-appointed interior.

ROVER VITESSE — £15,465

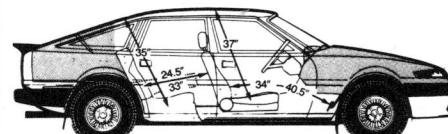

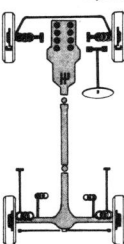

Price £	15,465
Capacity, cc	3528
Power, bhp/rpm	190/5280
Torque, lb ft/rpm	220/4000
Max speed, mph	132.1
0-60 mph, sec	7.1
30-50 mph in 4th, sec	7.6
mph/1000 rpm	29.4
Overall mpg	20.1
Touring mpg	26.2
Weight, kg	1393
Drag coefficient Cd	0.36
Boot capacity m³	0.36

Length 185" Width 69.8" Front track 59"
Wheelbase 110.8" Height 53.5" Rear track 59"

The Vitesse gives outstanding performance married to excellent flexibility and acceptable economy. The handling is predictable and most enjoyable — despite overlight steering — and ride quality is adequate; the brakes rumble, but do not fade, when used hard. Otherwise much like any other Rover, with mediocre accommodation (though very good load capabilities) for its size, and improved finish. A splendid car, and quite good value.

VOLVO 760 TURBO — £14,850

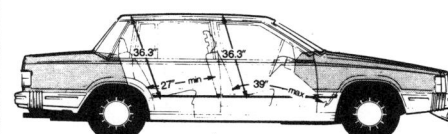

Price £	14,850
Capacity, cc	2316
Power, bhp/rpm	177/5700
Torque, lb ft/rpm	187/3400
Max speed, mph	115.8
0-60 mph, sec	9.3
30-50 mph in 4th, sec	7.0
mph/1000 rpm	25.2
Overall mpg	20.0
Touring mpg	26.8
Weight, kg	1338
Drag coefficient Cd	0.39
Boot capacity m³	0.40

Length 188.5" Width 69.3" Front track 57.9"
Wheelbase 109" Height 56.8" Rear track 57.9"

A 2.3-litre turbocharged "four" powers Volvo's executive flagship. Performance and economy are disappointing — little better than the V6 engined 760 GLE Auto — though subjectively the engine is smooth, willing and well suited to the car. Ride and handling are good, especially considering the Volvo's live axle. Roomy interior marred by poor driver's legroom. Merits are leather trim, good finish, plenty of equipment and a high build quality.

BMW's M-powered Coupé
— only for a few

WHEN motorsports enthusiasts think of BMW, images of the 3-litre CSL (lightweight) road car and its turbocharged derivative handled by the likes of Ronnie Peterson, Hans Stuck and Gunnar Nilsson spring to mind. Interest waned a little when the 635 was introduced in 1978, extra weight and superior comfort being acceptable in the market place but robbing the Coupé of its sporting image. Now a version, the M635 CSi, brings back the sparkle to BMW's Coupé and allows a fortunate few to vie with Porsche's 928S on the unrestricted *autobahnen*.

M-power is BMW's slogan today, fuelled by the success of the 3-series 4-cylinder block which is the heart of Nelson Piquet's Brabham Formula 1 car. The Motorsport division, headed by Dieter Stappaert and with designer Paul Rosche taking care of the technical developments, has already produced a high performing M535i and had all the equipment to hand to make a new flagship for the 6-series — a coupé style which has won the European Touring Car Championship more often than not over the

past ten years, and a 24-valve engine straight from the M1 Coupé, a model of which only 450 examples were produced, and a good one today is worth upwards of £50,000.

Looking back, you could say that the peak of the Coupé's development was in 1977 when the Motorsport division, then headed by Jochen Neerpasch, turbocharged the then 3.2-litre straight-six engine to produce between 750 and 800 horsepower, depending on boost. This machine's début

was at the Silverstone 6-Hours, and though the transmission was unequal to its task Peterson's handling of the monster was to make an indelible mark on our memory.

Neerpasch said then that it would take a year to produce a transmission that would stand up to the job, but by the time that year had elapsed he was embarked on a new course, to produce the stunning M1 Coupé in collaboration with Lamborghini, Ital Design and Baur. Though the M1 never reached the production levels required for homologation (until the FIA relaxed its

DEVELOPED for the M1 Coupé, the four valve six-cylinder engine now boasts even more power.

requirements) it resulted in the Grand Prix supporting M1 Championship rounds, which we remember as much as anything for the glorious noise of the 277 bhp, four-valve power units.

That power unit is the inheritance received by the M635, improved in fact by the adoption of Bosch Motronic fuel injection and digital ignition system, a higher 10.6:1 compression, and an increase in power to 285 bhp. Coupled with this increase in power is a substantial improvement in the torque curve, which now peaks at 251 lb ft at 4,500 rpm.

Coupling this power unit to the rear wheels, via a 25% ZF limited slip differential, is a superb new 5-speed close ratio gearbox which has reverse gear out on a limb to the left, fifth on a limb to the right. BBS split-rim alloy wheels carrying Michelin TRX tyres, 220/55 VR specification, take care of contact with the road, and the external appearance is completed with a deep chin spoiler and a discreet rubber spoiler across the bootlid.

By courtesy of BMW (GB) Ltd in Bracknell and the parent company in Munich we were able to spend a day with the M635 CSi in Germany in mid-January. Just seeing the car at the Frankfurt Show had made us determined to drive it as soon as possible, the car in our photographs being a pre-production model, series production starting about now for no more than 200-300 European customers. In 1985 production will be extended to 600-700, and this figure will include some right-hand drive models though the M635 will certainly remain very exclusive indeed.

The BMW Coupé has a classical shape, and a scarlet example still manages to look stunning on a bleak, otherwise cheerless day. We were advised to head south towards the Austrian border, on a lightly trafficked road to Garmish, so that 80 kilometres of autobahn could be covered as an initiation. The deep, expensive looking Recaro seats felt right straight away, and offered us a nice arm's length position behind the leather-covered steering wheel. There are just two main dials in the binnacle, a 280 kph speedometer and a rev-counter which has a thin red line starting at 6,500 rpm, thickening at 6,800 rpm where the rev-limiter operates.

We blessed BMW for continuing to exploit the normally aspirated potential of the engine, not the turbocharging route that they looked at in 1977 (and, indeed, as far back as 1969 when the 2002 Turbo won its class in the European Touring Car Championship). Turbocharging may be one expedient for extracting a lot of power from a comparatively small engine, but unless it's very well engineered it can leave the unit lacking in torque, and will certainly mute what could be an interesting engine / exhaust note. Not that we advocate noisy exhaust systems, but a well-tuned and suitably subdued note is, surely, part of the

pleasure of owning a sporting car?

Our first impression of the M635 CSi was that of a machine that could do credit to itself on a race track in standard trim. The driving position, the road feel that comes through the seats and the steering wheel, the well-tuned engine noise and the close-ratio gearbox all give that sensation of driving a competition car, a thoroughbred. Do not deduce, though, that the Coupé loses its refinement. The build quality remains excellent, the Karmann body going down

BODY alterations are minimal — a new chin spoiler and the tricolour M-Power badges.

the Dingolfing assembly line in the usual way, and the nice sound does not imply raucousness. The suspension feels firm, more at low speed, but by no means uncomfortable.

Though snow lay deep along the sides of the roads, the surface was well cleared and dry as we sped south, revelling in the BMW's performance. The engineers' performance chart confirmed our own impressions, first gear taking the car to 68

kph (42 mph), second to 110 kph (68 mph), third to 165 kph (102 mph), and fourth to 210 kph (130 mph). Other cars, seemingly reversing towards us at speed, made us slow down when the speedometer was hovering towards 240 kph (149 mph) in fifth, but the needle was still moving upwards at the time and there was no doubt at all that the Coupé would exceed 150 mph given time.

The 635 was rock-steady at these speeds, and felt completely undramatic too. It's a nice feeling that one can drive like this legally in Germany, though there is no way that we would overtake even one car, let alone two in convoy, at such a speed. But to slow down to a relaxing, comfortable and acceptable 100 mph is now a revelation to us, making us reflect on our tedious 70 mph motorway limit. And though totally in love with the refined howl of the power unit, we imagine it would become tiring after a long period at sustained high speed, meaning much above 100 mph.

On the other side of the coin, the M635

CSi is perfectly tractable right down to 40 kph, or 1,000 rpm, in fifth gear. We did plenty of that sort of motoring too, in the villages such as Oberammergau, home of the Passion play, and on icy roads which, though gritted, were not liked by the BMW on what the Germans scathingly call "summer tyres". "But *of course* it doesn't like snow," they said when we got back. "If we had known you were going into the mountains we would have fitted snow tyres." Well, even if we had known that winter tyres were on our options list we would have declined.

We were glad of the standard ABS braking system on the mountain roads, giving the BMW that extra dimension of safety. It is a peculiar experience to be able to brake hard on packed snow, and feel the pedal pulsating beneath your shoe. The car slows only gently, but in a straight line and under perfect control, this control remaining even if you turn the wheel.

Despite the power system the steering is nicely weighted, heavier than on a normal 635 by virtue of the wider tyres and passing on more feel of the road. As for the handling, we can only pass the opinion that it felt well-balanced and secure, since our distance of 340 kilometres in a day was all on motorways or snow-covered by-ways.

Comparisons will inevitably be made with Porsche's 928S 5-speed, since the two models have similar appeal and will be aiming for the same type of customer. Many customers, most perhaps, are totally loyal to one marque or the other and would not dream of switching, but still there will be many with an open mind in the market of a high-performance two-plus-two.

The Porsche, which has aluminium doors and engine cover, has a better power-to-weight ratio (310 bhp / 1,450 kg) than the BMW M635 CSi (285 bhp / 1,500

LEATHER-COVERED steering wheel complete with M-Power flash rounds off the interior.

kg), yet the performances are very comparable up to 100 mph, both reaching this mark in about 14.5 seconds. Beyond that the Porsche would edge ahead to a maximum speed of 164 mph, the BMW running out of puff at a claimed 158 mph. These maxima are, of course, purely academic even in Germany, but one can never deny the importance of having a higher top speed than the opposition!

Ironically, bearing in mind Porsche's forté of building sports / GT cars and BMW's mission of building sporting saloons, the Munich product is the more sporting of the two. The BMW is noticeably more taut, the seating and driving position more sporting, the engine note more enticing and the gearbox having closer-matched ratios. In cost terms, the BMW is five per cent dearer in Germany, and a similar disparity in Britain would take the price to £32,250 in the UK next year (at current values).

We understand that the M-power version of the 635 will not be homologated for competitions, since the 24-valve head would not be accepted as an evolution. That being so, a target of 5,000 identical cars being produced in 12 months is clearly out of the question. So, this is a model which will remain exclusive, for discerning customers, and the only question we cannot answer is why it took BMW so long to produce the M635 CSi. — M.L.C.

M TWICE

BMW's M for Motorsport saloon and coupe are heavy metal machines in the 140 and 150 mph class for cultivated speed. We test a truly dynamic duo, priced at £50,145 the pair!

Although the BMW M635CSi coupe and M535i are newcomers to Britain and less than a year old in German production, their roots lie in the seventies.

The M635 updates the 1978 M1 mid–engine power plant via the latest electronics and mounts all 286 hale and hearty horses in the usual BMW slanted manner, up front and driving the fat back wheels.

The basic six-series coupe shape, rated just under 0.40Cd in this guise, debuted in 1976 and is visually most familiar in 635CSi trim, when it costs over £26,100 and is powered by the 3430cc/218 bhp engine that you will find beneath the angular outline of our second test M-machine, the M535i. If you think the four door saloon lines look a little dated, body kit or no body kit, the reason is simply that the line dates back to 1972, with a 1981 update that also slipped the body just below 0.40Cd.

Unlike the M635, which was new to the market last year, the M535 was available in a previous BMW body, the 1979-81 production span establishing a high reputation for the cocktail of 3.5 litre six and 5-series space. The idea originally came from the separate, but still Munich-based, BMW Motorsport GmbH, who used to give VIPs such as their contracted drivers very special BMW hybrids to assess.

Complete with the no cost optional body kit, without which extended plastic panels the model is simply 535i, limited slip differential and the no-cost choice of three gearboxes, M535i provides genuine 140 mph pace for £17,950. Only the electric sunroof and Pioneer in-car entertainment were extras on the test car, which came with a close ratio gearbox to assist excellent 0-60mph times, marginally over seven seconds.

The silver M635 may not look very different to its £6000 cheaper 635CSi cousin, but we found it a classic high performance thoroughbred that actually does earn its price tag in Porsche 928S territory.

Of course the near 300 horsepower output of M635's unique 24-valve engine makes it a fast car, nearly 150 mph fast on our banked circuit and capable of leaping to 60 mph in an average 6.26 seconds. Yet that was not what captivated our testers. For this current coupe has the engine note of a Ferrari taught to sing Bavarian baritone, plus handling that takes the car power-slide from hooligan excess to a new art form.

Interior and controls

Both BMWs share the basics of all BMW cabins, clear white on black instrumentation and controls set out around a fascia moulded to sweep around the driver. Yet, the finishing touches and equipment levels were so different that only the constant presence of the same three spoke Motorsport steering wheel in both cabins underlined common parentage, and floorpan, for stylish coupe and upright saloon.

The M635 certainly smelt right! Some £856 spent on extra "Executive Leather" brought a pervasive whiff of class within. Unfortunately the seats themselves and their leather finish did not prove as effective in locating the occupants as the high-sided sports numbers in the considerably cheaper M535.

The coupe also has electrically operated front seat memory for three pre-set positions. Since the seats adjust in every conceivable plane this can be very useful for trips such as the one the editor did on return from Geneva, constantly swopping with assorted co-drivers.

In everyday use our testers found the three side adjustments, plus fore and aft with the in-out length selection that is possible on the steering column, perfectly sufficient on the M535. A good, commanding, driving position is a feature of the 5-series, whereas the Coupe has more the Mustang/Capri long-bonnet air in its lowline seating.

The large and matching dials of the M-machines differ only in detail; the coupe offers a readout to the equivalent of 280 kph or 180 mph and the beginning of the red warning zone at 6500. M535's saloon speaks of a 160 mph maximum and begins its rpm restriction at just over 6000 rpm; in fact the limiter allowed us 'only' 6000, or not quite that figure during performance testing.

Minor dials on the saloon include an inset to readout fuel economy; it appears you can have 20 mpg at all constant speeds up to 100 mph, which is not a bad achievement for 3.4 litres. As ever BMW do not use oil pressure gauges, providing water temperature and a fuel contents dial, the coupe's moving almost as fast as the car under provocation!

Both BMWs shared the MkII generation of computer which has the advantage of using a steering column stalk command as well as the usual pocket calculator layout. In the coupe the display called for a visit to a dealer when asked to talk about average mpg/mph, but the M535's was fully operational and the mpg figure displayed at the close of the test (21.5 mpg) was very close to our brim-to-brim calculations. Whatever BMW say about the new computers being easier to use than before, they still supplied a one page press briefing on how it should be used! Since some readouts demand up to *seven* depressions of the stalk, we think BMW should go back to the drawing board for a MkIII that does not require so much valuable driving concentration to be wasted.

Another area in which too much driver effort is frequently demanded is that of In Car Entertainment. The coupe came complete with a simple but excellent Blaupunkt Melbourne MR23 whilst the Pioneer AutoReverse with integrated graphic equaliser took longer to figure out. Even at this price level, BMW do *not* include the cost of ICE in their products.

Although we only used it occasionally, the standard air conditioning installed on the coupe proved notably effective; we did not measure its effect on fuel consumption, reeling already at sub-17 mpg figures returned by our intense enjoyment of M635 motoring...

Electronics are expected to serve the BMW driver today and both cars had test boards of warning lights for items such as light bulb failures, service interval indicators and electric motors for side glass and twin mirror operation. The extent of BMW's infatuation with servant sparks is best observed under-bonnet, where banks of fuses and five relay boxes lie clearly visible beneath transparent lid.

Cockpit contrasts continued via push-pull switchgear for the lights on coupe, and a combination of rockers and push-buttons for Hella headlamps and diminuitive auxiliaries in the M535 saloon.

However the most important changes between the two lay in the five speed gearboxes, and were immediately apparent to the driver...

Driving the M-cars

Installed behind the sturdy and carefully shaped steering wheel, the differences between BMW's Mean Machines are startling.

The coupe starts with a clean whirr that settles into a Bosch-monitored 700 rpm symphony of twin overhead camshafts, 24 valves and a 3.5 litre capacity. This enlarged capacity originally developed by Motorsport engineers for the legendary 1973 CSL Batmobile racing coupe.

The M535 is less dramatic with its conventional 3.4 litres, single overhead cam and 12-valves, but the test car was actually harder to drive!

This was not because of the engine. As usual the 3.4 litres – shared with 635CSi and 735i – is one of the most civilised and obedient engines you can buy. Unmatched amongst straight sixes for its combination of torque/useful horsepower/smoothness and economy.

The M535 problem in daily town use was simply that 'competition' layout for the close ratio gearbox, particularly when the transmission oil was cold and the first to second change 'across the gate' was at its most obstructive. It seemed a strange gearbox choice for the saloon with its comparatively mild 63.6 bhp per litre output hardly needing the 'on-the-boil' assistance of close ratio gearing.

By contrast this was the first M635 we had driven in RHD and the first with a

Top right engine bay: The M535 has a completely standard 3.5 litre straight six. Next to it is a wheel with the centre 'anti theft' cap removed. **Centre right engine bay:** In contrast the mighty 24 valve M635 engine with the M Power legend emblazoned on the cam cover. Next to it, the BBS wheel of the M635 has similar protection.

normal H-pattern for the first four speeds and fifth on an isolated plane, to the right of the H-pattern shift. Our previous experience had been in LHD cars equipped with Getrag gearboxes and shift patterns as for the test M535.

This makes more sense in that the twin cam/24 valve six does like to rev; peak power is developed 1300 revs above that of M535 and the output per litre is little lower than a TWR racing Group A Rover at 82.8 bhp.

However all our testers, who amassed 1511 miles on the demonstrator and have been fortunate enough to experience three other M635 sessions, agreed that they preferred the RHD choice of a conventionally laid out five speed. Since all are exceptionally keen drivers, one can understand that the racing pattern shift is just too much of a hassle in suburban use, introducing a degree of jerkiness that we found repeated even amongst BMW's Munich staff.

In road use the M535i is 95 per cent pure pleasure and remarkable value by BMW standards. For the men of Munich offer you Audi quattro-busting performance at comparable mpg, for a purchase saving measured in thousands of pounds. The troubles start when you try and extract the last five per cent! Generally this five-series M-car tracks along sibilantly and predictably. The test car's short gearing meant that 3500 to 4000 rpm could sound a little fussier than usual. The compensation was instant acceleration in any gear; fourth and fifth on M535 shattering the more powerful M635's times in the useful 50-70

mph band, the saloon returning 5.9/8.2 seconds where the coupe could manage only 8.3/11.5 seconds.

Naturally, when you get to a test track and start using the speeds the law does not permit in the UK the Coupe proves why it costs so much more with wheel–spinning performance that makes even the M535 look mundane in outright get up and go, although it was interesting to see that absolute top speeds were really not that different in a back-to-back session.

Thus, in everyday use, we were more than satisfied with M535's useable performance and pleasantly surprised by the comfort levels achieved, particularly in seating and a ride that quickly becomes compliant above town speeds, with very little of Renault's "tough on the streets" feedback coming through below 30 mph either.

Yet we still returned the M535 with an overall sense of disappointment. The dissatisfaction lay in the handling, or more particularly in the lack of adhesion displayed in either wet conditions, or when trying to extract that last five per cent spoken of earlier.

Driven in a relaxed gentleman's tourer manner, the M535 and its 220/55 VR Michelins support every move promptly. BMW's always seem eager to turn into corners, and M535 is no exception; it is just that leaving a bend can be so untidy and ragged!

For the M535 seems to settle to within 95 per cent of its capabilities easily enough, but when the throttle is fully depressed even the limited slip ZF

differential and much-vaunted sports package suspension cannot restrain tail slides that literally do vary from mild to wild.

The M635 has the benefit of more comprehensive Motorsport chassis changes, including a unique rear anti-roll bar layout and supplementary helper springs within the front struts, plus wider and lower Michelins (240/45 TRX).

The contrast between the two cars at the test track, and on a wet country journey was painful. Where the M535 floundered for grip, nose running wide followed by lurching tail-slide, the M635 cornered exactly where the driver pointed it, flattering ability so comprehensively that it just never felt as though it would spin, even when most of the view ahead could only be registered through the tinted side glass.

The M coupe achieved all this with a very creditable ride quality, slightly better than you find on the normal 635CSi.

However, there were some glaring quality failings on the M635 that we had not expected. Fundamentals such as the righthand wiper lifting off the screen at speeds well below the M-coupe's capabilities, and that old BMW trait of the side glass providing no seal with surrounding rubbers. This meant you could suffer whistles and squeaks from 60 mph, or that the glass would suddenly lift away at 100 mph. When Jaguar made their seventies coupe (XJC, not XJS), they put a lot of effort into beating this coupe side glass problem and BMW should by now have equalled that achievement.

When it came to the braking tests – unusually severe in that we tackle ten stops from 70 and 100 mph – the coupe's thoroughbred approach, with M1 four piston calipers assisting huge vented discs, shone through. Occasionally the M635 would veer slightly right, but otherwise its ABS system carried out its duties conscientiously and enjoyably during testing and subsequent road use.

Such was not the case for the M535. Average stopping distances from 100 mph were 44 *feet* extra, and 70 mph extracted a 12ft penalty for the saloon too.

In driving summary we came to realise that the M635 actually does deserve that worrying price premium over its already expensive bretheren. The M535 is at a bargain level, by the standards of the £20,000 category, because its somewhat dated lines and converted car feel to braking and chassis really do not justify nose-to nose pricing with the Audi quattro turbos, Mercs and Porsches, although there is a bit more to say about that in our summary.

Engine and performance

Take one look into the M635's engine bay and you know you are for an automotive treat. The M Power logo sits authoritively along with the 24-6 proclamation to distinguish the number of valves and cylinders.

Left: The top three pictures show the interior of the Five Series car, while the lower three that of Six series. This is what you get for all that money. Our M635 had the optional, ultra high quality finish, leather seats. There is an 'ordinary' finish leather seat available as a no cost option. You can also have the sport type seat (as fitted to our M535) finished in leather if you so desire.

The installation of the 24 valve six into the front of a BMW saloon, rather than a purpose-built mid-engined car such as its M1 ancestor, posed problems of space and mounting angle. For the original engine in M1 stood vertically and had dry sump lubrication. The motor we see in the front of M635 could not afford the luxury of vertical inclination because it would have protruded through the bonnet! So they installed the engine at a slant, and decided to use as many production 735/635 features as possible; thus the water pump is from series production and the usual BMW wet sump lubrication is employed.

Naturally the M1 engine did without the sophisticated electronic management that BMW customers take for granted today, so Robert Bosch Motronic was added to the modern recipe.

There were changes to the cylinder head waterways and inlet manifolding to suit that 30 degree slant installation, but the cylinder head casting and its 37mm/32mm inlet and exhaust valve dimension remain the same as for the M1 legend, along with the 264 degree duration camshafts that BMW F1 "father" Paul Rosche designed originally for M1.

Compared with the seventies legend in M1 the M635 motor has 286 bhp where once was 277. More importantly the torque curve has benefited with 246 lb ft at 4500 revs instead of M1's 244 lb ft on 5000 revs.

Such details were important, for our Mr Bingham demonstrated that BMW's claims of surprising economy for more than 1.5 tons of 3.4 litre coupe were perfectly possible. Philip covered 598 miles at an average 19.28 mpg, reaching to over 21 mpg in the Cotswolds and boosting our overall average from 16.98 mpg to 18.96.

However, you do not buy M635 for economy potential, and it was at the test track that its sheer speed shone through. Standing starts with tyres melting and the rear end slewing entertained, but dropping from a 4000 rpm clutch engagement to a mere 2500 introduced Herr Bavarian Batmobile. A character capable of reaching the best part of 100 mph in a quarter mile.

Where the M535 is quietly competent, munching up 0-100 mph in a shade over 20 seconds and hitting the drag strip quarter miler a hundredth under 15.5 seconds, arriving at under 90 mph, the 24-valver goes for it. To the accompaniment of glorious mechanical songs, stirring enough for any national anthem amongst a race famous for engineering, M-Coupe batters past 100 mph in under 17 seconds. It mashes the quarter mile barrier in just 15.89 seconds, speeding to a 95.5 mph average terminal velocity.

As we noted earlier the M535 saloon frequently beats big brother in fourth and fifth gear pulling trials, and its absolute best of a 141.1 mph flying quarter mile stood up well to the M635's 147.5 mph, recorded on the same gusty morning. For the man who likes a good performance per £ spent ratio, there can be no doubt that M535 at nearly half the cost of the test M635, is a bargain in BMW terms. Its suave engine, preferably mated to another gearbox, will also score over the coupe's scorcher in its suburban responses and

CONTINUED ON PAGE 41

BMW M535i TEST DATA

PERFORMANCE TEST RESULTS

All tests with a crew of two and a full tank of fuel.

Through the gears:

0-30mph	2.2secs	0-70mph	9.5secs
0-40mph	3.9secs	0-80mph	12.1secs
0-50mph	5.3secs	0-90mph	16.0secs
0-60mph	7.1secs	0-100mph	20.4secs

STANDING ¼ MILE:	15.5secs
TERMINAL SPEED	88.5mph
AVE TOP SPEED BANKED CIRCUIT	139.42mph
FASTEST ¼ MILE BANKED CIRCUIT	141.06

ACCELERATION IN 4th/5th:

30-50mph	6.0/7.9secs	60-80mph	6.1/8.5secs
40-60mph	5.9/8.0secs	70-90mph	6.8/9.1secs
50-70mph	5.93/8.2secs	80-100mph	8.3/9.9secs
90-110mph	11.8/11.7		

MINIMUM SPEED COMFORTABLE PULL AWAY:

4th	8mph	5th	13mph

MAXIMUM SPEED IN GEARS @ 6,000rpm:

FIRST	39mph	FOURTH	112mph
SECOND	60mph	FIFTH	141mph @
THIRD	81mph		

OVERALL FUEL CONSUMPTION:	21.07mpg
GOVERMENT FIGURES:	34mpg @ 56mpg; 28.5mpg @ 75mph; 17.7mpg Urban driving
PROVING GROUND FUEL CONSUMPTION:	8.94mpg
0-110mph	26.5secs

BRAKE TEST
10 STOPS FROM 100mph

MAXIMUM BRAKING EFFORT:		10 STOPS FROM 70mph	
AVERAGE	440ft		213ft
WORST	461ft [Stop 4]		228ft [Stop 7]
BEST	423ft [Stop 2]		195ft [Stop 1]

Testing carried out by Performance Car staff at the Motor Industry Research Association Proving Grounds, Lindley, Warwickshire and Millbrook Proving Ground, Luton, Beds.
Sun rolling road chassis dynamometer facility provided by Auto Technique, Unit C, Kingsway Industrial Estate, Kingsway, Luton, Beds. Tel: 0582 414000
Radio equipment supplied by Citizen Systems Ltd., London. Tel: 01-852 4607.

TRACK CONDITION:	Dry
TEMPERATURE:	+3°C
WIND SPEED	Ave: 8mph Peak gust, 27mph
BAROMETRIC PRESSURE	735millibars

SPECIFICATION

ENGINE TYPE:	Inline six cylinder, front-mounted
DISPLACEMENT:	3430cc
BORE:	92mm
STROKE:	86mm
COMPRESSION RATIO:	10:1
MAX QUOTED POWER (DIN):	218bhp @ 5200rpm
MAX QUOTED TORQUE (DIN):	224lbs/ft @ 4000rpm
BHP PER LITRE:	63.6
POWER TO WEIGHT RATIO (UNLADEN WEIGHT):	160.3bhp per ton
POWER TO WEIGHT RATIO (TEST WEIGHT):	138.8bhp per ton
FUEL SYSTEM:	ME-Bosch Motronic injection/ignition

CYLINDERS:	Iron block, seven crankshaft bearings, inline six
CYLINDER HEAD:	Aluminium, SOHC, 12 valves (2 per cyl)
GEARBOX:	Manual close ratio 5-sp, front mounted

GEAR RATIOS:

TOP	1.00	2nd	2.40
4th	1.26	1st	3.72
3rd	1.77	REVERSE	
CLUTCH		Hydraulic operation, single plate	
FINAL DRIVE RATIO:	3.07:1		

FRONT SUSPENSION:	MacPherson strut, BMW double link layout, plus anti-roll bar
REAR SUSPENSION:	Trailing arms (13°), supplementary linkage, coil springs, telescopic dampers and anti-roll bar
BRAKES:	11.2 inch discs, vented front
WHEELS & TYRES:	165 TR390 alloys + 220/55VR Michelin TRX
UNLADEN WEIGHT:	1390kg/3058lbs
TEST WEIGHT. CREW & EQUIPMENT:	1599.5kg/3519lbs
WHEELBASE:	103.4in
TURNING CIRCLE:	33ft
FUEL TANK CAPACITY:	15.4 galls
BASIC PRICE (INC TAX):	£17,950

OPTIONAL EXTRAS FITTED TO TEST CAR:
Electric sunroof, £635 + Pioneer ICE on test

BMW M635 CSi TEST DATA

PERFORMANCE TEST RESULTS

All tests with a crew of two and a full tank of fuel.

Through the gears:

0-30mph	2.3secs	0-70mph	8.7secs
0-40mph	3.7secs	0-80mph	10.8secs
0-50mph	4.8secs	0-90mph	13.2secs
0-60mph	6.3secs	0-100mph	16.7secs

STANDING ¼ MILE:	14.9secs
TERMINAL SPEED	95.5mph
AVE TOP SPEED BANKED CIRCUIT	144.5mph
FASTEST ¼ MILE BANKED CIRCUIT	147.5

ACCELERATION IN 4th/5th:

30-50mph	8.01/12.13secs	60-80mph	7.0/11.3secs
40-60mph	7.57/11.89secs	70-90mph	7.1/14.2secs
50-70mph	8.32/11.52secs	80-100mph	7.7/12.5secs

MINIMUM SPEED COMFORTABLE PULL AWAY:

4th	15mph	5th	17mph

MAXIMUM SPEED IN GEARS @ 6,500rpm:

FIRST	37.2mph	FOURTH	129mph
SECOND	63.0mph	FIFTH	148mph @ 6224rpm
THIRD	96.0mph		

OVERALL FUEL CONSUMPTION:	18.96mpg
GOVERMENT FIGURES:	34.5mpg @ 56mpg; 27.7mpg @ 75mph; 17.1mpg Urban driving
PROVING GROUND FUEL CONSUMPTION:	9.0mpg
0-110mph	21.5secs	90-110mph	8.7/11.9secs
0-120mph	26.9secs		

BRAKE TEST
10 STOPS FROM 100mph

MAXIMUM BRAKING EFFORT:		10 STOPS FROM 70mph	
AVERAGE	396ft		198ft
WORST	408ft [Stop 3]		206ft [Stop 8]
BEST	385ft [Stop 7]		183ft [Stop 9]

Testing carried out by Performance Car staff at the Motor Industry Research Association Proving Grounds, Lindley, Warwickshire and Millbrook Proving Ground, Luton, Beds.
Sun rolling road chassis dynamometer facility provided by Auto Technique, Unit C, Kingsway Industrial Estate, Kingsway, Luton, Beds. Tel: 0582 414000
Radio equipment supplied by Citizen Systems Ltd., London. Tel: 01-852 4607.

TRACK CONDITION:	Dry
TEMPERATURE:	+3°C
WIND SPEED	Ave: 8mph Peak gust, 27mph
BAROMETRIC PRESSURE	735millibars

SPECIFICATION

ENGINE TYPE:	Inline six cylinder, front-mounted
DISPLACEMENT:	3453cc
BORE:	93.4mm
STROKE:	84.0mm
COMPRESSION RATIO:	10.5:1
MAX QUOTED POWER (DIN):	286bhp @ 6500rpm
MAX QUOTED TORQUE (DIN):	246lbs/ft @ 4500rpm
BHP PER LITRE:	82.8
POWER TO WEIGHT RATIO (UNLADEN WEIGHT):	194bhp per ton
POWER TO WEIGHT RATIO (TEST WEIGHT):	169bhp per ton
FUEL SYSTEM:	Bosch ML-Jetronic

CYLINDERS:	Iron block, 7 bearing crank, 6 inline cyls.
CYLINDER HEAD:	Aluminium, DOHC, 24 valves (4 per cyl)
GEARBOX:	Manual 5-Sp, front-mounted

GEAR RATIOS:

TOP	0.81	2nd	2.08
4th	1.00	1st	3.51
3rd	1.35	REVERSE	3.71:1
CLUTCH		Hydraulic operation, single plate	
FINAL DRIVE RATIO:	3.73		

FRONT SUSPENSION:	MacPherson strut, Bilstein gas-damping + anti-roll bar and supplementary springs
REAR SUSPENSION:	Trailing arms (13°), coil springs, anti-roll bar and Bilstein gas-damping
BRAKES:	11.8in vented discs (F) and 10.7in vented (R); Servo-assisted
WHEELS & TYRES:	BBS 210 TR 415, 3-piece alloys + 240/45 VR TRX Michelin
UNLADEN WEIGHT:	1500kg/3300lbs
TEST WEIGHT. CREW & EQUIPMENT:	1720kg/3783lbs
WHEELBASE:	103.4in
TURNING CIRCLE:	38ft
FUEL TANK CAPACITY:	15.4 galls
BASIC PRICE (INC TAX):	£32,195

OPTIONAL EXTRAS FITTED TO TEST CAR:
"Executive leather" trim £856.00 + Blaupunkt Melbourne MR23 ICE on test

Don't be deceived by the innocent looks. BMW's new M5 is surely the ultimate Q-Car, a Plain Jane saloon with raving supercar performance. **Phillip Bingham** was one of the first to drive it.

Oh, the perfect picture of innocence! You wouldn't even notice it walking by. After all, it's only another boxy, booted saloon car, all dressed down with nowhere to go. It wears nothing more dashing than a shallow chin bib and a conservatively tidy metallic coat. No pronounced wheel arches, no striking coach lines, not even a hint of the rear spoiler found on all fast motor cars these days. No shouting or clamouring for attention whatsoever. Just another Five Series. Worse, probably one of those jumped-up, breathless tax beaters, a BMW 518.

But don't simply walk past. Pause for a moment and look again. Otherwise, you might miss the ride of your life. Didn't your Mother ever tell you, never underestimate the Plain Jane?

Aren't those wheels light alloy? And aren't they wrapped in chunky 220/55 VR390 Michelin TRXs? An 1800cc Five Series wouldn't know what to do with those.

And then there's the badges, ever so discreet but nevertheless there, to be found on the front grille and the boot lid. They show a letter 'M', with a tricolor. Surely it can't be a BMW Motorsport department identification? The same badge which was carried on Nelson Piquet's Brabham when he won the 1983 World Championship?

It is. BMW's racing department has once again been allowed to improve the breed. What can so easily be dismissed without so much as a second glance is one of the fastest true four-seater saloon cars in the world. A smug, understated facade in which the driver knows better. Around town, this BMW driver can enjoy absolute anonymity. You would hardly notice him. But out on the open road, the BMW offers anonymity of a very different kind. You'll be lucky if you can see it for dust.

They call it the BMW M5, the 'M' denoting the Motorsport department which builds it. Open the bonnet, and you'll find a derivative of the punchy 3,453cc in-line-six which powered BMW's beautiful but brief-lived M1 Coupe (*Performance Car,* March). The two main differences are these: the engine is now located up front, and it's even more powerful. The low-slung M1 Coupe was motivated by 277bhp; in its latest form, the 24 valve, 93.4 x 84mm unit turns out 286bhp at 6,500rpm. In other words, it's the same engine as that employed in BMW's 150mph 635CSi — but in this case the car is lighter and faster.

How fast? Hold on to your 286 horses, and in a moment I'll tell you. But don't let us miss the point in all this excitement.

Remember: at a price, there are quite a selection of road cars with this sort of power, or more. Many of them are lower, sleeker, prettier, faster. When you dare use them. But the truth of the matter is this: a car which is very obviously a streak of motorway lightning might as well be an Olympic sprinter in clown's shoes. It's so flash it can trip itself up. A loud confession of guilt whether crime has been committed or not. An open invitation for every *Flash Harry* to give chase and every passing policeman to look twice.

The understated Q-car goes a long way towards solving those problems. And you'll have to look long and hard to find any that are more understated than the BMW M5. Were it not for those two little 'M' badges, it would have perfected the Art of Deception.

Some time ago, the German performance car driver grew out of The Badge Mentality. One only has to observe the cars which come sweeping past on the *autobahn*: many leave an anonymous boot lid in their wake. Rather than add boastful badges, the Germans *remove* them. BMW recognise this fact. In their own words, they also recognise that "People are trying their best to change their cars, to individualise them, often in real performance rather than what only *looks* like the car's performance. So why should not BMW Motorsport do the same expertly?"

Of the 180 experts employed in the prestigious BMW Motorsport department in Munich, no less than 50 have contributed to the M5; as many as you will find working on the Formula One engine effort. According to the Head of Research and Development in the M department, Gerhard Richter, those 50 people also work on the M635, the forthcoming M3, and — think carefully about those models — a secret Group A project which should materialise next year.

BMW are adamant, however, that the M5 is *not* merely the fortunate by-product of an imminent competition car. "It is not meant for homologation purposes" says Press and Public Relations chief Michael Schimpke. "The M5 was built in order to demonstrate what BMW is able to do, and also to satisfy the individual requests of customers."

The last comment is endorsed by the fact that some 25 cars were delivered to customers *before* the M5 was formally introduced, at the Amsterdam Show in February. The Motorsport department has the capacity to produce only 250 copies of the M5 each year, and at £23,000 apiece there is already a twelve month waiting list in Germany. There are no plans at present to build right hand versions since, as Herr Richter points out, "Our problem is not with the engineering of the car, but the capacity to build it."

Tantalising, Schimpke adds, "Naturally we looked at Germany for a start, because it is our biggest market. That will take approximately a year or so. But then we will start thinking about other marketing opportunities."

The thinking behind the M5 itself appears disarmingly simple. Richter explains: "We have the four-cylinder engine, which in its highest form is in the

Formula One and Two cars. This is for the new M3. Then we have the six cylinder engine which we give the type number 88. You can already buy this in the M635, and the engine bay of the BMW 323 is just not big enough to take it. So the next obvious body for this engine was the Five Series.

"The M5 is a new combination of parts you can already find in other BMWs. But in the M5, several parts are hand made for high quality standard. Each M5's engine and suspension must be carefully checked and approved before it can leave our workshops."

The M5 differs from its closest relative, the M535i, in several important ways. Not least, there's that latent fireball stowed furtively beneath the bonnet. Before, 218bhp from the M535i was a perfect source of contentment. But now we've been spoilt. Let all those 24 valves go to work in the M5, and at a screaming 6,500rpm you are armed with 286bhp with which to combat distance. That makes the M5 a true four-seater with muscle to match, say, a mid-price Ferrari, and it's not far off Porsche's new 944 Turbo. Whereas the M535i is capable of 144mph, the more modest shape of the M5 is likely to come past you at anything up to 152mph. Our test drive comprised a mixture of ground-level flying on autobahns, and some dog-fighting with the twists and turns of Tyrolean mountain roads. Although the last 15mph crawls a little reluctantly on to the clock, the M5 will pull to 140mph vigorously, and sits at that speed with perfect, stable serenity.

The M5 spoils the driver on lazy days too. Its torque is the sort used to pull down tenement blocks; up from 244 lb.ft at 4,000rpm in the M535i to 250 lb.ft at 4,500rpm. That means its rarely *necessary* to change gear. If it were not for traffic lights, you could drive from one side of town to the other in fourth.

Not that the M5 driver is very likely to stoop to such sloppy ways. He won't be able to resist exploiting the firm, nickety precision of the five-speed Getrag gearbox. The first four gears catapult the saloon to an indicated 39, 67, and 100mph. Fifth can be left untouched until 135mph, driver happy in the knowledge that innocent looking Plain Janes may frequently misbehave without being noticed.

In a hurry, shifts must be made as near as dared to 6,500rpm. This is the point at which all 286 horses start pulling — but it's also where the tachometer needle must stop climbing. If it attempts to go any further, it will be slapped firmly about the face by an automatic rev limiter.

Fifth gear shows the signs of a guilty conscience after all the energetic activity encouraged by its predecessing ratios: as an appeasement to fuel economy, it is set at a relatively tall 0.81:1. Our 160 mile run returned approximately 23mpg. Quite reasonable, given that the homeward journey was covered at a pace more commonly associated with light aeroplanes.

A comfortable compromise between speed and serenity appears to be about 5,000rpm, or 125mph. In Britain, it would be necessary to keep a wary eye on the mirrors at anything beyond 2,900rpm.

At the other end of the velocity scale, BMW's M5 is equally dynamic. There is rarely reason to disbelieve the official claims made in Munich, so take it for granted that 60mph may be reached in 6.3 seconds, and a standing kilometre swept behind in 27 seconds. You'd have to go out and buy a Ferrari or top-end Porsche to improve upon that. Acceleration is similar to exotics such as De Tomaso's Pantera, Ferrari's Dino 308GT4 or Berlinetta Boxer 512, and Porsche's 928S. Think about the space and convenience provided by the M5, and then think about the others . . .

Just as important, the M5 is remarkably vigorous throughout the range of overtaking speeds. Without so much as changing out of fourth gear, this boxy saloon — with its unspectacular shape and 0.37cd — will accelerate from 63 to 88mph in a mere six seconds. Seven seconds later, a further 25mph will show on the clock.

At 100mph, we found the speedometer to be eight per cent optimistic. But don't think for a minute that this does anything to dull the sensations or driving satisfaction. The fact is, one mile of tarmac can still be relegated to the memory in 24 seconds. And at such a speed you'll still be able to hear the two 150 watt speakers beating out from their hiding place beneath the rear parcel shelf.

It is reassuring to know that this family-carrying missile can be returned efficiently to the level of Escorts and earthlings with the aid of ABS. Sensibly, this is a standard fitment. So, too, are beefed up springs and dampers, highly effective anti-dive, and discs all round with ventilations on the front pair.

Yet there are occasions when the M5 can be expected to — and *will* — show signs of brutishness. Axle tramp discourages cops-and-robbers type getaways, and in any of the initial three gears power oversteer may predictably be provoked if the engine is hovering in the power range. There is no reasonable right to expect anything else from a car which offers

203bhp for each ton.

But let us say this. When the scenery has merged to a multi-colour blur and the road is throwing a succession of twists and turns in your face, the M5 does a lot to help. And that's far more than many so-called supercars. High speed stability is so fine that you'll drive for miles at 140mph before so much as thinking about it. The steering, with variable power assistance and a nicely sized three-spoke wheel, is light yet informative. The brakes, in our limited test, were beyond reproach. The ride is firm enough yet also sufficiently unobtrusive, the clutch action just as you would find it any other saloon car. Body roll is modest, the inclination of those wide tyres to be diverted by camber changes no worse than should be expected.

By packaging its supercar performance in saloon car format, the M5 possesses that subjective but significant asset, driveability. Drive it quickly or drive it gently, life behind the wheel of the M5 is easy. Driver confidence is fully inflated within a few miles of introduction, visibility is unimpaired, the interior spacious, ventilation good, eardrums left untickled, palms remain perfectly dry.

The privileged few who frequently drive supercars might argue that the M5 lacks raw character, that it is *too* easy to make friends with, that there is no more challenge in eating up the road than there is in an arcade video game. To a certain extent, that might be true.

But is it not also true that speed need not necessarily be at the expense of a constant assault on the senses? After all, we live in civilised times. There are more raw sensations to be derived from a Chipmunk than from Concorde.

So let's not fool ourselves. Let's not go on pretending that the low-slung supercar is without doubt a superior means of conveyance. Quite apart from the uniformed attention overt supercars so readily attract, there's the continual threat of resentment and aggravation from other road users. Yet the BMW M5 is clever enough to deceive bystanders. It offers nearly all of the performance of many so-called supercars with very little of the bother. Quite literally, it may speed from one place to another before trouble shooters have even noticed it. When it is parked outside at night, its owner may sleep. When it is faced with a long journey, its passengers may reach their destination in unspoilt comfort. And when it is shown a clear road ahead, this impish devil wearing the smock of an Angel will simply melt into a heat haze on the horizon.

CONTINUED FROM PAGE 38

economy that totalled fractionally over 21 mpg for us; an overdrive manual or automatic box would be likely to show one or two mpg over that, which we think very reasonable for a genuine 140 mph saloon with space for four.

Summary

Our test and the pounds restively awaiting release from a bulging bank account, will decide which of the two M options you select, but how do the two cars compare against outsiders?

The obvious comparison for the £30,000 plus M635 is Porsche's excellent 928S at £35,524. You get much more of a sporting coupe with attendant diminished carrying capacity for luggage and people, but there is a 4.7 litre V8 currently offering 310 bhp, nearly 160 mph in our hands and very similar acceleration to M635 with slightly greedier appetite for petrol. Equally you might opt for the much cheaper, more civilised, but far less exciting Jaguar XJS V12 at £22,385.

That British torrent of cylinders and smooth 299 bhp stays over seven seconds to 60 mph because only an auto is fitted. There are two alternatives; go to Tom Walkinshaw Racing and have one of their five speed V12s, as tried by PC last year, and pay over £30,000 – or go back to Jaguar and have their 3.6 litre 24 valve six. The latter comes with a five speed Getrag at a list price of £19,248, or a convertible (almost unobtainable until in-house production has made inroads into the backlog) at more like £21,000.

Although sentiment and some enjoyable miles at the wheel of 3.6 cabriolet and coupe from Jaguar might make us complete a Coventry order form, Jaguar's 225 bhp is not enough to rival the BMW's similar four valve per cylinder six at present. Neither performance (0-60 mph, 7.2's; 0-100,21s; 139.8 mph max) nor reliable, thrilling, horsepower match BMW's 286 bhp.

So M635 is likely to find serious opposition coming mainly from fellow countrymen at Porsche, or possibly those

seduced by the V8 charms of Mercedes 500 SEC/500 SL convertible-coupe at £34,965 and £27,435 respectively.

The M535, with or without body kit, is harder to compare. BMW tells us that the body kits are not expected into production until "mid-Summer" and those confused by the more recently announced cocktail of 286 bhp 24-valver in 5-Series (M5) will find that BMW GB do not intend to offer such a car in RHD, "certainly not this year, and probably not in 1986 either."

Considering the 535's UK market rivals we find nothing at less than £20,000 to counter its four door pace/price package. You might buy an Alpina B10 from Frank Sytner (£25,348) or Audi's 200 front drive turbo of 182 bhp/140 mph capability for £19,998. The quattro 182 bhp turbo we tested recently costs £23,043 and is strongly recommended over the front drive model.

BMW plan to import 200 of the M635 and 100 are pre-sold with the UK management currently looking to secure another 50. No, we don't know where they all get the money from either, but if we find out you can be sure we would also buy BMW in this class for the combination of excitement and useable space.

The M535/535i was nothing like so convincing, but we do not think BMW will have trouble selling the 500 allocated at present, for BMW punters are used to paying over £16,000 for 528iSE.

Jeremy Walton

M1, M3, M5: it's a strange irony that BMW should have named its fastest road cars after British motorways blighted by a 70 mph speed limit. "M" in BMW-speak, of course, means "Motorsport", the division of the Bavarian car maker that goes racing and, in 1978, created the superb but short-lived M1 supercar in an attempt to clean up in Group 4/5. Alas, it never did, though the exposure of fame was assured with the invention of the M1 Procar series, a spectacular one-make supercar Challenge for racing drivers that warmed up the crowds before World Championship Grand Prix rounds in the early '80s.

The attrition rate in those races was formidable. But the M1 survived as a landmark: not just because it was very nearly the best car BMW never made — the whole project was jeopardised when build-partners Lamborghini pulled out at the prototype stage — but because the technology embodied in the M1's fabulous 24-valve twin-cam "six" sired a whole range of superfast M-cars. The deceptively-named M535i, incidentally, is not one of them, nor has it ever been. Both in series 1 and 2 guises it goes down in the record books as a factory shoe-horn special, mating the standard 3.5-litre single overhead cam straight six with the standard 5-Series bodyshell. With these cars, the Motorsport influence was purely cosmetic.

The M635 CSi we test here, however, was both designed and developed by the Motorsport divsion, even though it rolls off BMW's regular production lines. The M5 and M3 — neither of which is yet on sale in the UK — are built by Motorsport just down the road from BMW's HQ in Munich. Launched in 1983, some two years after the M1's demise, the M635 (since nicknamed the M6) inherited a modified version of the M1's 3453 cc engine. Digital Bosch Motronic fuel injection replaced the mechanical system and small modifications to the twin-cam, four-valves-per-cylinder head improved breathing efficiency. The result: maximum output raised by nine horses to 286 bhp at 6500 rpm and peak torque up from 239 lb ft at 5000 rpm to 251 lb ft at 4500 rpm. Set against the 218 bhp and 229 lb ft of the cooking 635 CSi, the M-car's performance potential is on a higher plane altogether.

Not that there are many visual clues. To say that the M6 is understated is itself an understatement. For a car that has assumed the M1's supercar status and, with a £34,000 price tag, is pitched right in the laps of Porsche, Ferrari, Jaguar and Mercedes, it might well be argued that fine-lattice style BBS alloy wheels with wide Michelin TRX tyres, a slightly deeper nose

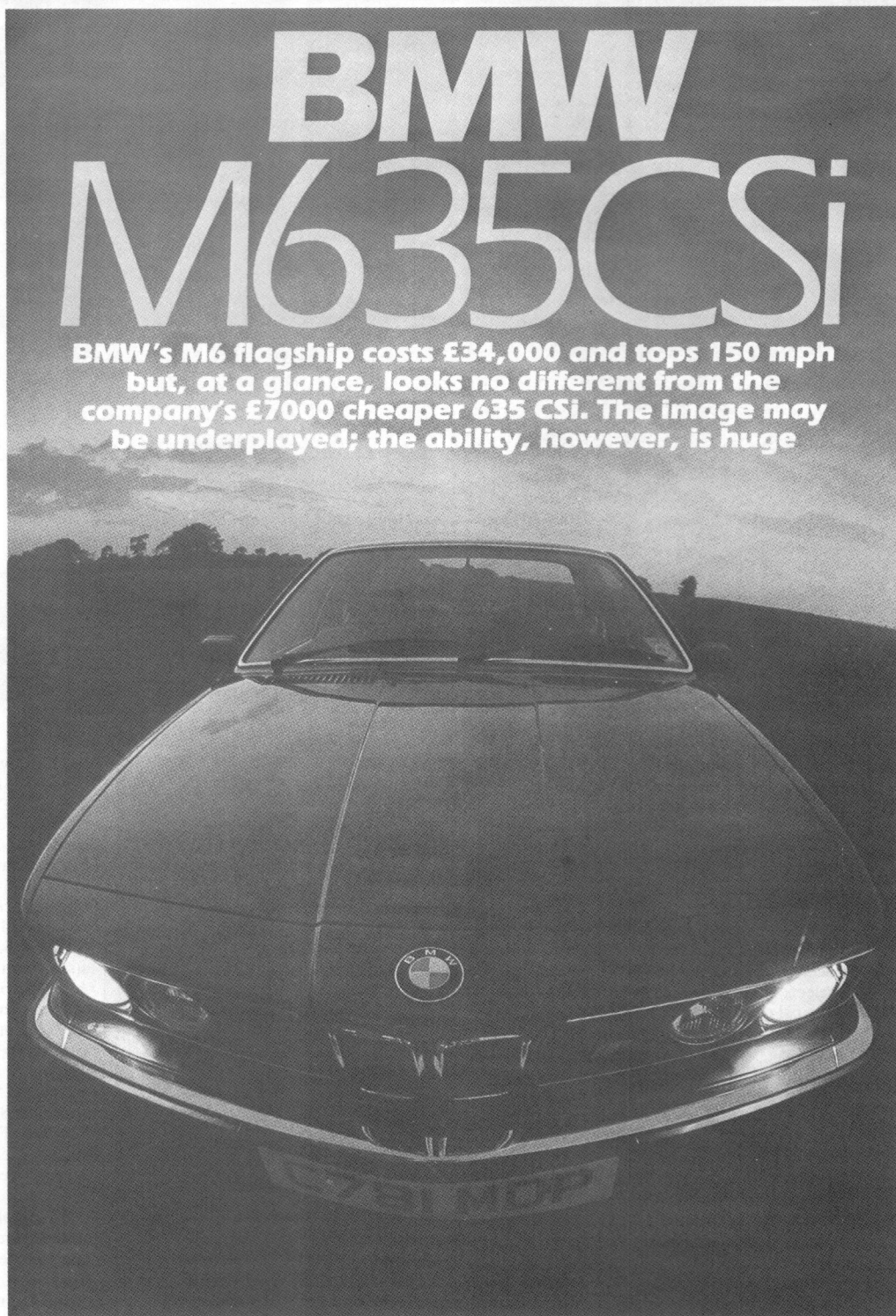

BMW M635CSi

BMW's M6 flagship costs £34,000 and tops 150 mph but, at a glance, looks no different from the company's £7000 cheaper 635 CSi. The image may be underplayed; the ability, however, is huge

spoiler, colour-keyed door mirrors and two oh-so-subtle grille and boot-mounted Motorsport emblems provide insufficient visual distinction — too much sheep and not enough wolf. Then again, the cognoscenti *will* know that the M6 is capable of well over 150 mph and of accelerating from rest to 60 mph in a shade over 6 seconds. When you've got it, you don't necessa-

rily have to flaunt it. Those inclined to dismiss a shape that's been around for the best part of a decade do so at their own peril.

We knew the M6 was Ferrari-fast when we strapped our fifth wheel to an early left-hand drive example in Germany and published our findings in *Motor* w/e November 12, 1983. At that time, there were no plans to make a rhd version for the UK,

but so strong was demand that BMW relented this spring, reporting firm orders for well over half of the 200 rhd cars earmarked for the UK market in '85. Back in '83 we predicted a price of £30,000 for the M6 if it ever reached these shores: the reality is a hefty £33,875, though this includes a long list of equipment not found as standard on the German-spec cars — the

extra-wide 210 mm BBS wheels and 240/45 VR 15 tyres and electrically powered sunroof, for instance. Even so, the M6 finds itself up against some redoubtable machinery, in the shape of other superfast two-plus-two coupés like the Mercedes 500 SEC (just superseded on the German market by the 560 SEC but still available in the UK at £37,065), the Jaguar XJ-S HE (nearly £10,000 cheaper at £23,995), the Audi Quattro (£23,273) the Ferrari Mondial Quattrovalvole (£33,200), and Porsche 928 S II (£35,524).

In view of the weight of ability and charisma stacked against it, we decided to re-test the M6 on our home ground. At Millbrook as in Germany, the M-car's performance was mighty though, as expected, the effects of tyre scrub knocked about 4 mph off the top speed, dropping it from 154.1 mph to 149.7 mph. On the basis that the autobahn figure is a true reflection of the BMW's top-end potential, however, it ranks with the very fastest cars in its class. Even comparing the BMW's Millbrook top speed with those of rivals tested at Millbrook, none is faster, though it has to be borne in mind that the Porsche (149.1 mph) was tested in automatic form: with a manual gearbox, it should just shade the M6. But such hair-splitting is academic: 150 mph is fast enough.

The BMW clearly has a surfeit of power over traction. Even with the best efforts of the fat Michelins, it left the line at Millbrook in a haze of tyre smoke to reach 60 mph in just 6.3 sec — 0.1 sec better than the lhd car managed in Germany — and on to 100 mph in an even more impressive 15.2 sec, a figure some 3.1 sec better than that for the cooking 635. Only a handful of the world's supercars are significantly quicker over these benchmarks though there is a dense concentration of machinery that gets close. Only fractions separate

the BMW from the Ferrari (6.4 sec), the Porsche (6.5 sec) and the Audi (6.5 sec) in the sprint to 60 mph, for instance. Few rivals, however, can match the sheer breadth of the BMW's power band: in fourth, it covers each of the 20 mph increments between 20 and 110 mph in 6½ sec or less. In fifth, the increment times between 20 and 90 mph are disposed of in around 9½ sec apiece: the 24-valver has real strength in depth.

And that's the way it feels on the road: it's one of the few engines that can combine a solid low-speed punch with genuine free-revving fury. Its power is impeccably delivered, too, with sabre-edged throttle response and a complete lack of temperament in traffic. Outstanding tractability is another facet of this remarkable engine, full throttle being accepted from just 600 rpm in fifth without a tremor of protest. The crowning glory, however, is aural. For a unit with

a 24-valve head, mechanical refinement is exceptional all the way to the 6900 rpm red-line. Nor has smoothness been compromised: in this respect, the M-car's engine is as silken as any BMW's. The difference is a subtle one, the well-bred hum that is a BMW trademark hardening and deepening with revs until full throttle unleashes a spine-tingling howl that, while far from loud, is delectable in its sophisticated quality.

Porsche, Lotus and others have shown that hard muscle and a gentle thirst aren't necessarily mutually exclusive. Inasmuch as our M6 returned just 16.9 mpg overall, then, it is rather old-fashioned. Even so, it's unlikely that potential customers will view this too gravely in the light of the BMW's storming performance. If pricked by barbs of social conscience, however, the customer might find that Ferrari's 18.6 mpg Mondial avoids sleepless nights. With a large

dose of restraint, our projected touring figure of 24.0 mpg is certainly on, giving a potential maximum range of 370 miles on a 70-litre tankful of 4-star.

The M6 has a heavy clutch but a delightfully light and easy gearchange. The five-speed Getrag gearbox is notable for its speed and precision and urges more frequent gear-changing than is strictly necessary. The intermediate ratios are long but closely spaced, allowing maxima of 38, 64, 99 and 133 mph at 6900 rpm. Second, in particular, is a superb overtaking gear, enabling the engine to deliver scintillating acceleration to just under the legal limit.

On first acquaintance, the big BMW's handling seems less than special. The assisted recirculating ball steering feels low-geared, a mite heavy and lifeless; initial turn-in isn't as crisp as the taut, nuggety low-speed ride might lead you to believe. Yes, the M6 even verges on the ponderous around town.

But then it was never made to be enjoyed in an urban environment. With increasing pace comes more articulate communication and lucid responses. Steering that seemed dead suddenly shows more than a glimmer of life: steady and smoothly coordinated hands are now required on the helm for jerky inputs are faithfully transmitted to the front wheels. There the fat Michelins bite hard, holding understeer tightly in check. Squeeze the throttle firmly and the M6 points more directly still. In fast sweeps, chassis balance is simply superb, the cornering attitude remaining neutral for far longer than with the cooking 635. The 25 per cent limited slip differential, uprated gas-filled dampers and tauter springs help tame the rear semi-trailing arms' natural predilection for oversteer, restricting camber angle changes under power and on lift-off. When the tail does drift out of line it is to the command of the throttle and in

BMW M635 CSi

PERFORMANCE

WEATHER CONDITIONS

Wind	10 mph
Temperature	58 deg F/14 deg C
Barometer	29.6 in Hg/1003 mbar
Surface	Dry tarmacadam

MAXIMUM SPEEDS

	mph	kph
Banked Circuit (5th gear)	149.7	240.9
Best ¼ mile (5th gear)	151.0	243.0
Terminal speeds:		
at ¼ mile	98	158
at kilometre	123	198
Speeds in gears (at 6000 rpm):		
1st	38	61
2nd	64	102
3rd	99	159
4th	133	214

ACCELERATION FROM REST

mph	sec	kph	sec
0-30	2.7	0-40	2.4
0-40	3.7	0-60	3.7
0-50	4.9	0-80	5.1
0-60	6.3	0-100	6.7
0-70	8.1	0-120	9.1
0-80	10.0	0-140	11.7
0-90	12.2	0-160	15.2
0-100	15.2	0-180	19.7
0-110	18.6	0-200	25.5
0-120	14.9		
Stand'g ¼	14.9	Stand'g km	26.6

ACCELERATION IN TOP

mph	sec	kph	sec
20-40	9.4	40-60	5.7
30-50	9.3	60-80	5.6
40-60	9.4	80-100	5.5
50-70	9.3	100-120	5.5
60-80	9.3	120-140	5.7
70-90	9.7	140-160	6.4
80-100	10.1		

ACCELERATION IN 4TH

mph	sec	kph	sec
20-40	6.7	40-60	4.2
30-50	6.7	60-80	4.1
40-60	6.6	80-100	3.8
50-70	6.1	100-120	3.7
60-80	6.0	120-140	3.5
70-90	6.1	140-160	3.7

80-100	6.2	160-180	4.2
90-110	6.6	180-200	5.6
100-120	7.8		

FUEL CONSUMPTION

Overall	16.9 mpg
	16.7 litres/100 km
Touring*	24.0 mpg
	11.8 litres/100 km
Govt tests	17.1 mpg (urban)
	34.4 mpg (56 mph)
	27.7 mpg (75 mph)
Fuel grade	97 octane
	4 star rating
Tank capacity	70 litres
	15.4 galls
Max range*	370 miles
	595 km
Test distance	517 miles
	832 km

*Based on official fuel economy figures — 50 per cent of urban cycle, plus 25 per cent of each of 56/75 mph consumptions

STEERING

Turning circle	10.8 m, 35.6 ft
Lock to lock	3.2 turns

NOISE

	dBA
30 mph	67
50 mph	70
70 mph	74
Maximum†	80

†Peak noise level under full-throttle acceleration in 2nd

SPEEDOMETER (mph)

True mph	30	40	50	60	70	80	90	100
Speedo	33	44	55	66	77	87	96	106

Distance recorder: 1.4 per cent fast

WEIGHT

	kg	cwt
Unladen weight*	1506	29.6
Weight as tested	1708	33.6

*No fuel

Performance tests carried out by *Motor's* staff at the Motor Industry Research Association proving ground, Lindley, and Millbrook proving ground, near Ampthill.

Test Data: World Copyright reserved. No reproduction in whole or part without written permission.

GENERAL SPECIFICATION

ENGINE

Cylinders	Six in-line
Capacity	3453 cc
Bore/stroke	93.4/84.0 mm
Max power	286 bhp 210 kW at 6500 rpm (DIN/ISO)
Max torque	251 lb ft 185 Nm at 4500 rpm (DIN/ISO)
Block	Cast iron
Head	Aluminium alloy
Cooling	Water
Valve gear	Dohc, 4 valves per cylinder
Compression	10.5:1
Fuel system	Bosch Motronic electronic fuel injection
Ignition	Fully programmed
Bearings	7 main

TRANSMISSION

Drive	To rear wheels
Type	5-speed, manual

Internal ratios and mph/1000 rpm

Top	0.81/23.8
4th	1.00/19.3
3d	1.35/14.3
2nd	2.08/9.3
1st	3.51/5.5
Rev	3.71:1
Final drive	3.73:1

AERODYNAMICS

Coef Cd	N/A

SUSPENSION

Front	Independent by Macpherson struts, lower wishbones, anti-roll bar; coil springs
Rear	Independent by semi-trailing arms, anti-roll bar; coil springs; limited slip differential

STEERING

Type	Recirculation ball
Assistance	Yes

BRAKES

Front	Ventilated discs, 30 cm dia
Rear	Discs, 28.4 cm dia
Servo	Yes
Circuit	Split diagonally
Rear valve	ABS anti-lock system

WHEELS/TYRES

Type	Aluminim alloy, three-piece 210 mm dia
Tyres	240/45 VR15 TRX
Pressures F/R (normal)	29/32 psi 2.0/2.2 bar
(full load/high speed)	32/43 psi 2.2/3.0 bar

ELECTRICAL

Battery	12V, 90 Ah
Alternator	80 Amp
Fuses	30
Headlights	
type	Halogen
dip	110 W total
main	220 W total

GUARANTEE

Duration	12 months unlimited mileage
Rust warranty	Six years

MAINTENANCE

Determined by "service indicator"

Make: BMW **Model:** M635 CSi **Country of Origin:** West Germany
Maker: Bayerische Motoren Werke AG, Munich
UK Concessionaire: BMW (GB) Ltd, Ellesfield Avenue, Bracknell, Berkshire RG12 4TA. Tel: (0344) 26565
Total Price: £33,875
Options: Air conditioning (£1457) **Extras** fitted to test car. Air conditioning

Above: the quintessential BMW facia — clean, understated, ergonomically correct, if conventional. Left: for clarity, instruments rank among the best

Above left: front seats are hard but properly supportive, adjustable for height, rake, reach and thigh support. Above: despite inviting looking buckets rear legroom is minimal. Left: one of the world's great production engines beautifully presented

accordance with the driver's wishes: it can be held or cancelled at will. Nor is the BMW's composure unduly upset by mid-corner bumps. As with the ride, control is the key — firm but never harsh. For sheer power, the brakes measure up to the high dynamic standards the M6 sets itself. For feel and progression, however, they disappoint, the pedal feeling dead and "wooden" under foot, characteristics BMW seems unable to get out of its ABS-equipped cars.

Mechanically, the M-car is no noisier than the ordinary 635 but its fatter tyres do rumble more over coarse surfaces. A pity, because wind noise is very low. For the rest, the M6 is almost pure 6-Series coupé, a car that has been featured numerous times in these pages. The interior is essentially roomy for two people, though tall drivers will run out of headroom, despite the height adjustment offered by the front seats; the rear accommodation is strictly "occasional". Instrumentation is comprehen-

sive and beautifully clear, switch-gear ergonomics every bit as good as the driving position and the relationship between the major controls. The M6, like its cheaper counterpart, has one of the best heating and ventilation systems in the business and retains the useful Service Interval Indicator and on-board computer. If you can come to terms with sitting in a cabin that looks and feels no different from that of the £7000 cheaper 635 CSi, it's easier to appreciate the first-class standard of build and finish. Much the same goes for the outside, too.

But if that bothers you then the M6 isn't your sort of car. It's a car that doesn't wear its mighty heart on its sleeve. The essence of the M6 is tangible rather than visible — its pace and dynamic virtuosity. It's too thirsty and, by some measure, too expensive. But genuine Grand Touring cars of the BMW's ability are becoming all too rare. For those who can appreciate what it has to offer, the price is merely an obstacle.

The Rivals

The M6 has only a handful of real rivals. We detail them below

BMW M635 CSi — £33,875

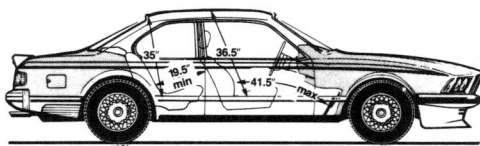

Length 187" Width 68" Front track 56"
Wheelbase 103" Height 54" Rear track 57.5"

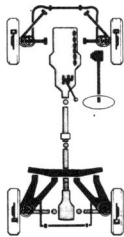

Capacity, cc	3453
Power, bhp/rpm	286/6500
Torque, lb ft/rpm	251/4500
Max speed, mph	149.7*
0-60 mph, sec	6.3
30-50 mph in 4th, sec	6.7
mph/1000 rpm	23.8
Overall mpg	16.9
Touring mpg	24.0
Weight, kg	1506
Drag coefficient Cd	N/A
Boot capacity m³	0.35

Fastest BMW available in the UK takes the decade-old 6-Series firmly into supercar territory and more than holds its own. Fabulous 24-valve six delivers stunning performance with good refinement but is very thirsty. Handling is both enjoyable and forgiving, ride firm but well-controlled. Interior is too ordinary for £34,000 but build and finish are first class. Generally well equipped but air conditioning is extra.

AUDI QUATTRO — £23,273

Length 173·3" Width 67·8" Front track 57·3"
Wheelbase 99·3" Height 53" Rear track 56"

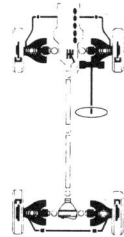

Capacity, cc	2144
Power, bhp/rpm	200/5500
Torque, lb ft/rpm	210/3500
Max speed, mph	138e
0-60 mph, sec	6.5
30-50 mph in 4th, sec	8.2
mph/1000 rpm	23.5
Overall mpg	19.9
Touring mpg	24.7
Weight, kg	1260
Drag coefficient Cd	0.43
Boot capacity m³	0.21

*Figures for old model

A more civilised machine these days than its epoch-making predecessor, the 4wd Quattro is more than ever a milestone in car design. It combines phenomenal roadholding and traction with performance, refinement, economy, comfort and accommodation in a way that has no equal, against which its weaknesses — poor ratios (still) and slow shift, unprogressive heating, sparse instruments — are minor failings. Now available with anti-lock braking.

FERRARI MONDIAL QV — £33,200

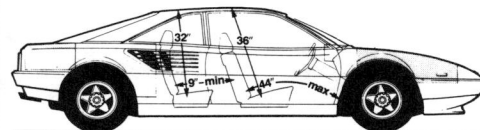

Length 180.3" Width 70.5" Front track 59.8"
Wheelbase 104.3" Height 49.5" Rear track 58.8"

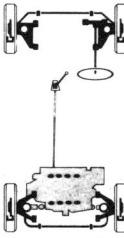

Capacity, cc	2927
Power, bhp/rpm	240/7000
Torque, lb ft/rpm	192/5000
Max speed, mph	146.1
0-60 mph, sec	6.4
30-50 mph in 4th, sec	5.6
mph/1000 rpm	19.7
Overall mpg	18.6
Touring mpg	18.6
Weight, kg	1448
Drag coefficient Cd	N/A
Boot capacity m³	N/A

The strictly two-seater 308GTB is a closer rival to the BMW on price, but the Mondial is a token four-seater. *Quattrovalvole* power delivers supercar performance with respectable economy, while other virtues include safe, balanced handling, superb brakes, a comfortable ride and driving position, smooth engine and satisfying gearchange. Low gearing leads to highish noise levels, however, and the heating and ventilation leave something to be desired.

JAGUAR XJS 5.3 HE — £23,995

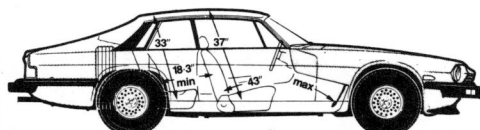

Length 191·8" Width 70·5" Front track 58·5"
Wheelbase 102" Height 49·8" Rear track 58"

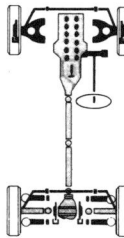

Capacity, cc	5345
Power, bhp/rpm	299/5500
Torque, lb ft/rpm	318/3000
Max speed, mph	152.4
0-60 mph, sec	7.5
30-50 mph in kickdown, sec	2.9
mph/1000 rpm	26.8
Overall mpg	16.3
Touring mpg	20.2
Weight, kg	1707
Drag coefficient Cd	0.40
Boot capacity m³	0.24

Available only in automatic form, and supplied with the May "Fireball" heads, the magnificent V12 XJ-S is now noticeably more economical. As always, it offers an astonishing combination of high performance and refinement. Its handling is less sporting than that of the BMW, but safe and predictable. Styling hasn't improved with familiarity, and there is not much room in the rear, but it remains one of the most desirable high-speed expresses.

MERCEDES 500 SEC — £37,065

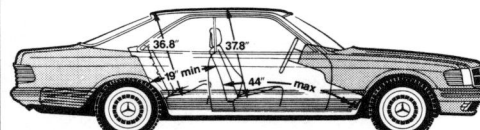

Length 193.3" Width 72" Front track 60.8"
Wheelbase 112" Height 55.3" Rear track 59.8"

Capacity, cc	4973
Power, bhp/rpm	231/4750
Torque, lb ft/rpm	299/3000
Max speed, mph	141.6
0-60 mph, sec	7.9
30-50 mph in kickdown, sec	2.9
mph/1000 rpm	32.9
Overall mpg	17.5
Touring mpg	23.3
Weight, kg	1636
Drag coefficient Cd	0.34
Boot capacity m³	0.44

Current flagship of the Mercedes range (New 560 SEC comes to UK next year), the 500 SEC is superbly engineered and impeccably finished. Good performance and handling is matched by excellent refinement, a very comfortable ride, and full four-seater accommodation. Only available with a four-speed automatic transmission which has exceptionally high overall gearing but is still responsive and smooth. Very economical for its size and performance.

PORSCHE 928 SII — £35,524

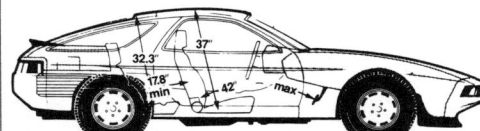

Length 175" Width 72·3" Front track 61"
Wheelbase 98·5" Height 51·8" Rear track 60·3"

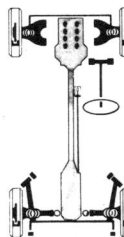

Capacity, cc	4664
Power, bhp/rpm	310/5900
Torque, lb ft/rpm	295/4100
Max speed, mph	149.1
0-60 mph, sec	6.5
30-50 mph in kickdown, sec	2.2
mph/1000 rpm	26.5
Overall mpg	17.1
Touring mpg	22.5
Weight, kg	1539
Drag coefficient Cd	0.38
Boot capacity m³	0.21

In its latest form Porsche's splendid 928S is stunningly quick and respectably economical with its new four-speed automatic transmission. Handling and road-holding superb, and potent brakes now have anti-lock system as standard. Very spacious and comfortable cabin for two (but a cramped 2 + 2), excellent driving position and instrumentation, but ride and refinement not in the Jaguar class. Manual option would be more in keeping with its sporting character.

H&B BMW M635CSi

The classic business suit bulges with more muscle

THE 6-SERIES BMW, in U.S.-legal form, has always been just a bit on the sedate, understated side, with less power than its substantial chassis and beautiful body would call for. The original 630CSi appeared in mid-1976; as a 1977 model for the U.S. it produced 176 bhp from an L-Jetronic 3.0-liter six. With emissions standards tightening, the extra 224 cc of the 633CSi gave it just one more horsepower in 1979 and another four by 1982, as BMW refined its U.S. plumbing. The 6-series was no slouch, but hard-driving BMW enthusiasts were tantalized by the European 635CSi, which had been putting out

218 *Pferdstärke* since 1978, not to mention the mid-engine M1 with a twin-overhead-camshaft, 24-valve version of the six producing 277 bhp. The latter car was brought in as a gray-market item for awhile; it's now history, but its formidable engine now sits at the front of the well-known coupe. Giving it, at last, the muscles to fill out the suit.

The 24-valve coupe is known as the M635CSi. M is for Motorsports, the *Rennabteilung* that has churned out Touring, Formula 2 and Formula 1 victories for the marque, as well as the M1 in its own special series. The 6 stands for the body type, of course, the 35 indicates the 3.5-liter (actually 3453-cc) displacement, and the Coupe-Sport-injected letters have been on the U.S. cars all along. But the engine is even stronger than the original M1's; with Bosch Motronic injection, a higher compression ratio (up to 10.0:1 from 9.0:1) and new induction and exhaust systems, this greatest of all contemporary inline-6s puts out a whopping 286 bhp DIN, equal to approximately 281 SAE, and an estimated 251 lb-ft of torque. Now that it has all the power a sane person can ask for, what does it do with it?

Just about everything a sane person can ask for. You probably already know that an M635CSi did 154.6 mph in our World's Fastest Cars exercise (R&T, September 1984). Our current test car, a 1984 model made EPA and DOT-legal by H&B of Berke-

ley, California, and further enhanced by H&B chassis, wheel, body and interior equipment, is an exhilarating performer. Once you get to 3000 rpm, the acceleration becomes explosive, yet the car's refinement allows it to charge downfield without ripping out a single shoulder seam. Clearly a matter of having your steak and eating it too. Stay off the throttle, and the M635CSi is the comfortable, if slightly stiff and bulky coupe you've always known; get on it and you'll have the full attention of every traffic enforcement officer in your state. In making the engine emissions-legal, H&B doesn't seem to have lost any of the power, although at the top—6000 rpm to redline—the six seems just a tiny bit less willing than the one we ran in Europe in 1984. The gearbox, a strengthened 5-speed, is mated to a 3.73:1 final drive (compared with 3.45 for the regular factory 635CSi) and gives better dig off the line if a slightly lower maximum speed. The shifting is notchy when the box is cold, but effortless after that.

H&B's chassis kit includes special springs front and rear, tuned to the factory-installed Bilstein shocks, and adjustable anti-roll bars. The wheels are 3-piece H&B 8 x 16-in. alloy, mounting 225/50VR-16 Yokohama A-008 tires. The body and interior options are mostly appearance oriented, although the extra gauges (oil pressure, oil temperature, voltmeter), four Hella quartz halogen headlights and aerodynamic side skirts are →

AT A GLANCE

	H&B BMW M635CSi	Ferrari Mondial Cabriolet	Porsche 928S
Price, base/ as tested	$48,207 $57,382	$65,000 $66,180	$50,000 $50,000
Curb weight, lb	3350	3545	3425
Engine/drive	inline-6/rwd	V-8/rwd	V-8/rwd
Transmission	5-sp M	5-sp M	5-sp M
0–60 mph, sec	6.4	7.6	5.9
Standing ¼ mi, sec @ mph	15.0 @ 95.5	16.0 @ 87.0	14.2 @ 101.5
Stopping distance from 60 mph, ft	141	153	145
Lateral acceleration, g	0.826	0.808	0.820
Slalom speed, mph	58.8	60.3	57.9
Fuel economy, mpg	16.5	13.5	est 17.0

	Pro	Con
M635 CSi:	responsive and powerful engine, excellent brakes, superb handling and room for four	high price, Teutonic interior appointments not for everybody, sparse parts availability
Mondial Cabriolet: tested 5-84	good overall performance, Italian charm, exciting sounds	limited rear-seat room, weight, modest acceleration in class, difficult folding top, price
928S: tested 4-85	excellent power and handling, high-quality finish, adequate luggage room	poor outward vision, weight, limited rear-seat room

ROAD & TRACK
R&T
ROAD TEST

ft, from 80 in 249, and with excellent ABS control at all times. With brakes like these, you can go incredibly fast with full confidence. The H&B suspension modifications give a tauter ride and less body roll, but without undue harshness.

The interior is all business, of course, in the usual BMW ultra-authoritative way, with fairly stiff seats (uncomfortably narrow for larger persons and shy on head room with the sunroof) and little design "warmth." That's okay if you're all business, but if that's the intent, then we can nitpick a bit: The steering wheel adjustment could be improved, the shift lever ought to be a little closer, and the instrument lighting could be better for those with older, less adapting eyesight. But the coupe's rational body style gives space and outward vision of the kind you just can't get in the mid-engine exotics with which you normally associate this level of performance.

In short, the H&B M635CSi is the BMW coupe that enthusiasts have been wanting for nearly nine years. When the 630CSi first arrived, we missed the agility of the preceding 3.0CS; now this bigger, tougher car has the power and handling to cope with any situation. It strongly impressed every member of the staff, including those of less competitive bent. After all, it can go slowly very well, too. Just not very often.

functional (at the extreme ends of the performance spectrum, at least). The result of all this equipment (besides adding more than $9000 to the already hefty $48,207 asking price) is superb comportment in all but the tightest places (parking lot jockeys may never know how good this particular Bimmer is). The lateral acceleration figure is 0.826g, slalom speed is 58.8 mph (remember, this a big car), and the brakes are absolutely splendid, pulling the 3485-lb car down to nought from 60 mph in just 141

PRICE

List price, San Francisco	$48,207
Price as tested	$57,382

Price as tested includes std equip. (elect. window lifts, trip computer, central door locking), H&B alloy wheels and Yokohama A-008 tires ($2716), air cond ($1484), H&B suspension group ($1469), elect. sunroof ($743), leather int ($679), AM/FM stereo/cassette ($599), metallic paint ($411), H&B instrument group ($394), security alarm ($222), striping ($216), elect. antenna ($149), tinted glass ($84), warning triangle ($9)

ENGINE

Type	dohc 4-valve inline-6
Bore x stroke, in./mm	3.68 x 3.31....93.4 x 84.0
Displacement, cu in./cc	211...........3453
Compression ratio	10.5:1
Bhp @ rpm, SAE net/kW	est 281/210 @ 6500
Torque @ rpm, lb-ft/Nm	est 251/340 @ 4500
Fuel injection	Bosch Motronic
Fuel requirement	premium unleaded, 92 pump octane

DRIVETRAIN

Transmission	5-sp manual
Gear ratios: 5th (0.81)	3.02:1
4th (1.00)	3.73:1
3rd (1.35)	5.04:1
2nd (2.08)	7.76:1
1st (3.51)	13.09:1
Final drive ratio	3.73:1

GENERAL

Curb weight, lb/kg	3350	1521
Test weight	3485	1582
Weight dist (with driver), f/r, %		55/45
Wheelbase, in./mm	103.3	2625
Track, front/rear	55.9/58.3	1420/1480
Length	187.2	4755
Width	67.9	1725
Height	53.3	1354
Trunk space, cu ft/liters	15.7	445
Fuel capacity, U.S. gal./liters	18.5	70

CHASSIS & BODY

Layout	front engine/rear drive
Body/frame	unit steel
Brake system	11.2-in. (284-mm) vented discs front, 11.2-in. (284-mm) discs rear; ABS; vacuum assisted
Wheels	cast alloy, 16 x 8
Tires	Yokohama A-008, 225/50VR-16
Steering type	recirculating ball, power assisted
Turns, lock-to-lock	3.5
Suspension, front/rear:	MacPherson struts, double-pivot lower links, coil springs, tube shocks, anti-roll bar/modified semi-trailing arms, coil springs, tube shocks, anti-roll bar

CALCULATED DATA

Lb/bhp (test weight)	12.4
Mph/1000 rpm (5th gear)	24.0
Engine revs/mi (60 mph)	2500
R&T steering index	1.35
Brake swept area, sq in./ton	220

ROAD TEST RESULTS

ACCELERATION

Time to distance, sec:

0–100 ft	3.2
0–500 ft	8.2
0–1320 ft (¼ mi)	15.0
Speed at end of ¼ mi, mph	95.5

Time to speed, sec:

0–30 mph	2.4
0–50 mph	4.9
0–60 mph	6.4
0–70 mph	8.5
0–80 mph	10.7
0–100 mph	16.9

HANDLING

Lateral accel, 100-ft radius, g	0.826
Speed thru 700-ft slalom, mph	58.8

FUEL ECONOMY

Normal driving, mpg	16.5

BRAKES

Minimum stopping distances, ft:

From 60 mph	141
From 80 mph	249
Control in panic stop	excellent
Pedal effort for 0.5g stop, lb	20
Fade: percent increase in pedal effort to maintain 0.5g deceleration in 6 stops from 60 mph	nil
Overall brake rating	excellent

SPEEDS IN GEARS

5th gear (6100 rpm)	est 145
4th (6500)	129
3rd (6500)	97
2nd (6500)	65
1st (6500)	38

INTERIOR NOISE

Constant 30 mph, dBA	63
50 mph	66
70 mph	75

ACCELERATION

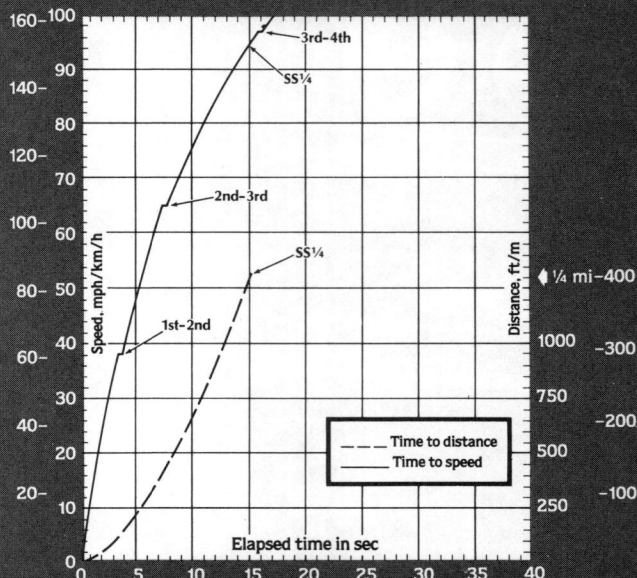

Unusually for BMW, engine is a four cylinder – but with a difference. Twin cams, 16 valves and 2.3 litres result in 200 bhp and 146 mph maximum speed for compact car

Traditionally – in West Germany at least – BMW has ruled the roost when it comes to sporting saloons. The Munich company's whole marketing philosophy has been centred around the performance aspects of its cars.

It must, therefore, have been a slap in the face for BMW when arch-rivals Mercedes-Benz launched the twin-cam Cosworth version of its 190E. Suddenly, the Mercedes-Benz 190E 2.3 16 was *the* stormer to be seen in.

Never ones to submit to such a challenge, BMW laid down plans to produce an 'M' version of its smallest car, the 3-series. And true to their reputation, they've built a stunningly fast compact car, even though it's strictly for the racing fraternity at present.

Heart of the car is a 2.3-litre version of BMW's straight-four two-litre engine, with a four valve/cylinder aluminium head related directly to the larger six cylinder engine. Maximum power is 200 bhp at 6750 rpm and 177 lbs ft torque is produced at 4750 rpm. A close ratio Getrag five-speed gearbox is used with first on the left dog-leg, opposite reverse.

Suspension at both ends is based on standard 3-series, but much modified. The strut type front suspension incorporates twin-tube gas shock absorbers and a thicker anti-roll bar than standard. Power steering is fitted as standard. The independent rear suspension utilises semi-trailing arms and spring struts and anti-roll bar. ABS brakes with ventilated discs at the front and solid at the rear are used.

Like all cars destined for the race track (the M3 is primarily a homologation special for Group A saloon car racing), careful attention has been paid to the cars' aerodynamics. The rear window angle has been steepened by three degrees and mates to a new plastic composite boot lid, 1.57 ins higher than the standard cars, surmounted by a wing made of the same material. A front spoiler-cum-bumper, metal wheelarch extensions and

THRILLING THREE

moulded extended sills help to reduce the standard car's Cd from 0.38 to 0.33.

Tank capacity has been increased to 15.4 gallons with claimed fuel consumption figures of: urban, 24.3 mpg; constant 56 mph, 48.7 mpg and constant 75 mph, 37.6 mpg.

Internally, BMW haven't skimped either. The driver and front passenger have the luxury of superb, body-hugging Motorsport seats.

The essence of a car like the BMW M3 is its performance. Its makers claim a maximum speed of 146 mph and a 0 to 62 mph time of 6.7 secs. Unfortunately, traffic conditions encountered during a brief test drive curtailed a maximum speed run, but an indicated 136 mph was easily attained and the engine felt as though the additional 10 mph would be achieved.

Particularly impressive at those high speeds was the car's stability: BMW's time spent developing the cars' aerodynamics and suspension has not been wasted. If the car does have a fault, then it becomes apparent during motorway cruising. Fifth gear is a direct top and not an overdrive, and

as a consequence the engine always feels slightly strained and uncomfortable while cruising.

Take the car off motorways and onto open, twisting roads and the M3 comes alive. It is extremely well balanced in all aspects. The engine pulls strongly from well down, gaining strength beyond the 4000 rpm mark and continuing right round to the redline in each gear until maximum in fifth is reached.

The handling is a delight, precise and vice-free. Admittedly, the rear wheels can be made to slide out of line, but the LSD helps to prevent any sudden breakaway and a touch of opposite lock will soon balance the car and have it pointing the right way.

A car with the M3's performance potential needs good brakes and the ABS-assisted discs perform exceptionally well: although if pressed too hard one will hear the familiar ABS knocking, just back off slightly and it disappears.

The driving position is near ideal, with the pedals perfectly located for heel-and-toeing. The car being left-

hand drive, its gearchange can catch the British out at times – especially from fourth to third – but practice makes perfect.

Left-hand drive? Unfortunately, yes. Such is the location of the engine and the convoluted exhaust manifold that there simply isn't room beneath the bonnet for a right-hand drive steering rack. Actually, that isn't strictly true. BMW engineers did admit that a right-hand drive model could be engineered, but the redesigned exhaust manifold would rob the engine of ten bhp and they couldn't allow that.

More's the pity, because the BMW M3 is a driver's car in every sense of the word and would certainly find a ready market in the UK.

MODEL:	BMW M3
RANGE	2-door saloon
ENGINE:	2.3 litre, 200 bhp
PERFORMANCE:	146 mph, 0-60 6.7 sec
MPG:	24.3 – 48.7
DATE IN UK:	No import likely

German autobahns breed fast cars and, in the £20,000 pric bracket, they don't come much faster than BMW's M535 and Mercedes 190E 2.3-16. Both give you a seaso ticket in the fast lane, but one has the edge . . .

What happens when a good heavyweight steps into the ring with a brilliant middleweight? Chances are the heavyweight belts the living daylights out of the middleweight. Agility and finesse, after all, only count for so much. In the end, weight of punch is usually the deciding factor.

It would be easy (and it's certainly tempting) to superimpose the "unfair match" analogy on the fight about to take place here. BMW's be-spoilered M535i, for instance, is a whole class size bigger than Mercedes' 190E 2.3-16. It's power-to-weight ratio, however, that determines how hard a car hits and, although the BMW's 3.5-litre straight-six delivers a mighty 218 bhp, it has to push around a bulky 1391 kg. The 185 bhp developed by the Mercedes' 2.3-litre 16-valve four-cylinder engine, on the other hand, is burdened with a much more modest 1258 kg. Do the appropriate sums and what emerges are surprisingly close power/weight ratios: 157 bhp/tonne for the BMW against 147 bhp/tonne for the Mercedes. Consider the influence of its class-leading Cd (0.32 against the BMW's 0.37) and the high-performance status of its £21,045 price tag and there's every reason to believe that it should fight the £19,495 M535i for the executive hot-rod title. And that's without mentioning its dazzling handling or legendary build-quality.

There are other parallels. Both cars' rear wheels are driven via five-speed gearboxes, steering is by power-assisted recirculating ball in each case, and disc brakes all round (ventilated at the front) are the order of the day. The BMW has coil-sprung all-independent suspension by MacPherson struts at the front and trailing arms at the rear with anti-roll bars at both ends and gas-filled dampers all round. The 2.3-16 has gas-filled dampers, too, with MacPherson struts and lower wishbones at the front and

POWERFUL MAGIC

from a standstill, the BMW reaches 60 mph in 6.9 sec (7.4 sec for the Mercedes) and 100 mph in 18.9 sec (20.7 sec). The superior torque of the BMW's bigger engine is even more decisive when it comes to fourth gear flexibility: the M535 takes 7.6 sec for the 30-50 mph increment and 7.5 sec for the 50-70 mph increment. More significantly, all the 20 mph increments between 20 and 80 mph are covered in just over 7 seconds apiece, pointing to an exceptionally flat torque curve.

Even with its much shorter gearing in fourth (17.8 against 23.5 mph/1000 rpm) the Mercedes can't pull out much of a lead over the BMW: over the 30-50 mph increment in fourth it's 0.2 sec adrift at 7.8 sec; 7.0 sec for 50-70 mph puts it only 0.5 sec ahead.

There's less in it all out, both cars just failing to crack 140 mph (139.1 mph for the BMW and 137.1 mph for the Mercedes). With slightly longer gearing than its 23.5 mph/1000 rpm in fourth, the BMW would probably go a little faster still since its top speed corresponds with 6025 rpm, some 800 rpm beyond peak power. The Mercedes, on the other hand, is close to being ideally geared for top speed, pulling slightly less than the 6200 rpm at which its engine develops peak power. On a long, downhill stretch of autobahn the 2.3-16, with its more slippery shape and 7000 rpm "red-line", would probably overhaul the brick-like M535, but the BMW would storm back on the next incline. Both can claim to be true 140 mph cars. But when the Sierra Cosworth goes on sale this spring, that won't be good enough.

Subjectively, the BMW gains even more ground over its four-cylinder rival. While far from the sweetest engine in the BMW armoury, the sohc 3.5-litre straight six is smooth enough to make even the best four-pot units seem like primitive industrial generators. Allied to the superb spread of power already illustrated and sparkling part-throttle response, you have civilised potency beyond this price level. Deficient in capacity and cylinders, the 2.3-16's engine has to work a great deal harder for its living. But, denied the M535's thumping torque, its very willingness to rev hard is a valid substitute for lazy flexibility. At around 4000 revs, the Cosworth-developed 16-valver comes alive, pulling with what seems like ever-

Mercedes' clever and sophisticated multi-link independent system at the rear, which also features hydropneumatic self-levelling to prevent excessive squat under hard acceleration. Like the BMW, anti-roll bars are used at both ends but only the Mercedes has a limited slip differential.

The Bosch KE Jetronic-injected and bored-out version of the 190E's M102-series four-cylinder engine, equipped with its 16-valve Cosworth head, develops 185 bhp at 6200 rpm and 173 lb ft of torque at 4500 rpm. With an extra couple of cylinders and 1.2 litres more capacity, the Bosch ME Jetronic-injected straight-six fitted to the M535i manages 218 bhp at 5200 rpm and 224 lb ft of torque at 4000 rpm. The torque is decisive.

PERFORMANCE

Whatever the paper statistics might predict, the Mercedes isn't in the same race as the BMW on the road. Its tighter dimensions and lighter weight are no substitute for the sheer torque of a big engine, and there's little this side of desperation that can shrink the gap on fast roads punctuated with pockets of slow-moving traffic: the BMW is off and running before the Mercedes has had time to take a deep breath. By most standards, the 2.3-16 is a rapid car but, in this confrontation, the M535 is boss.

The figures say it even more starkly. Smoking its squat Michelin TRX tyres expensively

BMW turns in crisply and holds a chosen line obediently

increasing vigour to 7000 revs. Power delivery is crisp, tractability exemplary. But the abiding character of the engine is almost racer-peaky and that means a whole lot of gearchanging to keep it on the boil. As 2.3-litre four-cylinder engines go, the Mercedes' is a minor masterpiece. The bottom line is that it's simply not enough engine for the money.

ECONOMY

Predictable glory for the Mercedes, though not by a crushing margin. The respective consumptions returned after nearly 300 miles of hard driving in the West country were 23.5 and 20.4 mpg. One might have anticipated a sub-20 mpg figure for the deep-chested BMW, but bear in mind that the BMW didn't have to work as hard to keep up with the Mercedes as the Mercedes did to keep up with it.

Asked only to maintain a brisk pace, our projected touring consumptions of 24.5 mpg for the BMW and 32.5 mpg for the Mercedes should be easily attainable, giving the Mercedes a significantly longer range than its Bavarian rival since both have 70-litre (15.4-gallon) tanks. The BMW's 377 miles is good by any standards, but the Mercedes' 500 miles is exceptional.

TRANSMISSION

Our original M535 test car (*Motor* w/e March 9, 1985) had ZF's optional close-ratio five-speeder with direct fifth giving 23.5 mph/1000 rpm (still longer than the 2.3-16's 22.5). The "overdrive" box fitted to this car, however, retains the close-ratio's direct gearing for fourth and makes fifth a remarkably long-striding 0.81:1 for 29.0 mph/1000. The first three gears are more widely spaced, too, with first shorter and second and third longer than

Mercedes has limpet-like grip and unflappable poise

before. Apart from more relaxed cruising, the changes do little to modify the car's forceful character on the road, which says much for the 3.5-litre unit's sheer flexibility. At 6000 rpm, maximum speeds in the first three gears run to 37, 64 and 101 mph with top speed achieved in fourth instead of fifth.

No such liberties have been taken with the Mercedes' ratios. They were chosen to extract plenty of "edge" from an engine that might all too easily be swamped by rangy gearing. In our view, Mercedes could have made the intermediates closer, especially between third and fourth, without seriously compromising refinement. As it is, the first four gears allow 38, 62, 89 and 125 mph at 7000 rpm.

Perhaps the biggest advantage of ordering the BMW with the "overdrive" transmission is its conventional gate pattern which, unlike the ZF's, has fifth and not first out on a dog-leg. Shift quality is superior, too, with an almost delicately slick and precise action for such a powerful car. The Mercedes's ZF-style Getrag shift is consistently quick when the occasion demands but lacks the finesse of its rival's in normal driving, proving a mite notchy, weakly sprung across-gate and baulky into the dog-leg first. Matters aren't helped by a clutch which bites very early and rather abruptly. The BMW's clutch,

while heavier, is more progressive.

HANDLING

It is fortunate for the BMW that our Twin Test was conducted in the dry. For as long as its fat Michelins can maintain adequate purchase of the tarmac-adam, the M535 handles with outstanding assurance and great balance. What happens when it rains, however, has long been an issue of some controversy between *Motor* and BMW. They say its sudden appetite for oversteer is an acceptable characteristic no more exaggerated than in any other BMW model. We say not, and have proved that the car's wet weather behaviour improves out of all recognition when wearing Pirelli rubber.

Dry roads, however, provided the M535 with ample opportunity to prove what a wieldy machine it can be, even on its standard TRX tyres. In spite of feeling a big car from behind the wheel, its responses are neat and secure, its steering quick and beautifully weighted. Grip is abundant with mild understeer in tight turns changing to a more neutral attitude with speed. The BMW turns in crisply and holds a chosen line obediently. Although severe bumps and dips can cause the

Both engines to drool over, but it's the BMW's 3.5-litre "six" (above) that packs the real wallop

body to pitch mildly when pressing on – firmer damping would almost certainly cure this – the M535's composure is never thrown. Oversteering antics in the dry require a brutal application of power in a low gear and a tight bend but, unlike the wet road variety, the resulting tail slide is slow and progressive, easy to catch or hold.

For all the BMW's dynamic prowess, its handling is eclipsed by that of the Mercedes which, as we have said on numerous occasions, is just about the best-handling saloon money can buy. The 2.3-16 has it all: limpet-like grip, unflappable poise, a helm capable of providing a finely shaded picture of the road's surface and an ultimately forgiving nature. Understeer is well contained, even in tight bends, and nothing less than a bomb crater could upset its rock-like stability. Body roll is greater than the BMW's but transient handling more fluid, the 2.3-16 weaving through a series of S-bends with lucid, tautly-controlled body movements. Deliberate provocation will unstick the tail in the wet, that much we know from previous experience, but in the dry it takes almost foolhardy excess. Even then the tail is unlikely to step out more than a few inches and a relaxed grip on the steering wheel is all that is required for its castor action to apply the correct amount of opposite lock. The steering's power assistance, like the BMW's, is almost perfectly judged but allied with a matchbox-crunching precision the Bavarian car lacks.

RIDE COMFORT

Initially, the BMW seems the more comfortable car with the better rounded ride. And it's true that around town the M535 takes the sting out of sharply edged ridges and troughs more effectively than the 2.3-16. But it soon becomes apparent that the Mercedes' suspension is superior at coping with just about every other type of surface irregularity. In particular, it is blessed with exceptional damping control that defies any hint of float or wallow at speed. It's the sort of ride you appreciate on a long run. The BMW's ride, while accomplished, is just a touch more flabby when worked hard though, on most roads, you'd hardly notice.

BRAKES

Another scalp for the Mercedes.

Above: Familiar BMW facia still looks good with fine instruments and ergonomics. Recaro front seats (right) offer superb support. Rear legroom (far right) is ample

If its brakes are slightly overservoed, it's a small price to pay for their tremendous bite, feel and stopping power. The BMW's brakes – equipped, like the Mercedes, with the ABS antilocking system as standard – also offer potentially life-saving security, but while undeniably powerful they feel mushy underfoot and, thus, inspire less confidence than they should.

ACCOMMODATION

No surprises here. The BMW is the bigger car and offers appreciably more room – both for passengers and for luggage. Although not especially spacious by, say, Ford Granada standards, the M535 can cater for two pairs of six-footers in tandem with enough rearward front seat travel and headroom for very tall drivers. Oddments space is generous, too, with a large glove box, rigid front door pockets, a large cubby in the centre console and a moulded tray on top of the facia.

The Mercedes lacks the cubby but compensates with elasticated pouches on the front seat backrests. In broader terms, however, it is severely handicapped by its lack of rear legroom, which is little better than that provided by most 2+2s. Mercedes don't see this as a problem, in the light of statistics that show most 2.3-16 customers to have been Porsche 944 refugees, but, to the best of our knowledge, the 190 was never designed as a 2+2 – it has four doors – and therefore shouldn't be judged as such.

Above: Mercedes facia is neat and businesslike but centre console gauges are hard to read at a glance. Bucket-style front seats (right) aren't as good as BMW's, rear room (far right) cramped

AT THE WHEEL

Both cars have good driving positions but it is the BMW that has the most comfortable, supportive seats. Recaros of generous dimensions and decent lateral and lumbar padding, they locate the driver with rare conviction. By comparison the Mercedes' Recaro-copies seem rather amorphous but, by normal standards, they are well shaped and a great improvement on the 190E's upholstered planks.

Fore/aft adjustment is generous in both cars, and both sets of front seats adjust for height as well. Unlike the Mercedes', the BMW's pedals aren't offset. The

Mercedes' single column stalk looks neat but is necessarily more complicated than the BMW's pair: understandable and easy to use supplementary switchgear is a feature common to both cars.

Easy to see out of and place on the road, both cars also have decent mirrors and headlights well up to their 140 mph performance potential.

INSTRUMENTS

The fact that the Mercedes has more dials than the BMW doesn't automatically make it the winner in this category. Although its trio of auxiliary gauges are easier to see than most of their console-

mounted ilk, you wouldn't see them in a BMW which, while maybe guilty of paucity, sets standards of clarity and presentation few can match. It also has a second generation on-board computer fitted as standard.

HEATING AND VENTILATION

Both cars have first-class systems with simple, smooth-acting controls governing powerful and versatile yet easily regulated heaters. The Mercedes' is slightly more sophisticated in having separate temperature controls dedicated to either side of the cabin, but the BMW's ventilation – independent like that of the

---

BMW M535i   £19,495

Make: BMW **Model:** M535i
Country of Origin: Germany
Maker: Bayerische Motören Werke AG, Munich
UK Concessionaire: BMW (GB) Ltd, Ellesfield Avenue, Bracknell, Berkshire RG12 4TA
Tel: (0344) 26565
Total Price: £19,495

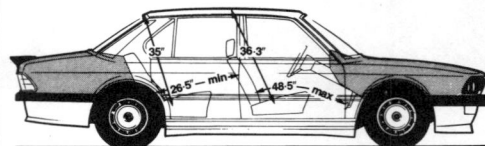
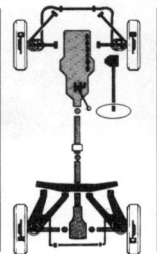

	35"	36·3"	
	26·5" – min	48·5" – max	
40%			56%

Length 4·62m (182") Width 1·70m (67") Front track 1·43m (56·3")
Wheelbase 2·62m (103·3") Height 1·29m (51") Rear track 1·46m (57·5")

MERCEDES-BENZ 190E 2.3-16 £21,045

Make: Mercedes-Benz **Model:** 190E 2.3-16
Country of Origin: Germany
Maker: Daimler-Benz AG, Stuttgart
UK Concessionaire: Mercedes-Benz (UK) Ltd, Delaware Drive, Tongwell, Milton Keynes MK15 8HA
Tel: (0908) 668899
Total Price: £21,045

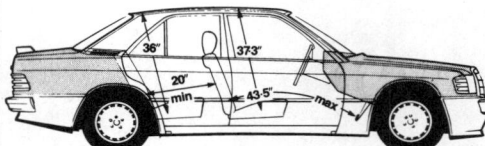
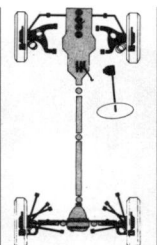

	36"	37·3"	
	20" min	43·5" – max	
46%			54%

Length 4·42m (174·3") Width 1·70m (67") Front track 1·44m (56·8")
Wheelbase 2·66m (105") Height 1·36m (53·5") Rear track 1·43m (56·3")

harshness even at high revs, though very definitely a straight-six. Road roar is a little more prominent than in the Mercedes, but wind noise just as low.

FINISH

Up against virtually any car but the Mercedes, the BMW would emerge victorious, here. Few cars are more tightly screwed together and assembly quality is consistently fine throughout. Under the sub-heading of "build-quality", though, the Mercedes is on a higher plane. The car just feels that much more solid; panel fit and paint quality are close to perfection.

It says something about Mercedes' confidence (arrogance?), then, that the 2.3-16's interior *looks* cheap and nasty with unattractive seats and shiny wood panels. It doesn't look £21,000-worth, it doesn't even look £15,000-worth. The M535, on the other hand, contrives to look smart and tasteful with materials that are almost certainly cheaper and less durable than the Mercedes'. It succeeds.

EQUIPMENT

The blob chart shows that the M535 *just* racks up more points, which is hardly encouraging for Mercedes whose slower, smaller car costs some £1150 more – £2395 if you order the BMW without the M-Technics body kit, front seats and suspension settings: the plain 535i.

Both cars have central locking, electric windows and door mirrors, headlamp wash-wipe systems, height and tilt adjustment for the front seats, alloy wheels and tinted glass. Items exclusive to the BMW include heated mirrors, auxiliary driving lamps and adjustable steering. The Mercedes bounces back with seat back map pockets, an illuminated vanity mirror and self-levelling rear suspension. Neither car has a radio.

CONCLUSIONS

Is the BMW simply too strong for the talented Mercedes? It looks that way. The 2.3-16 verges on true greatness in some areas – dynamically and aesthetically, certainly – and goes remarkably hard for a 2.3-litre car. The Mercedes is also built and finished to an exceptionally high standard and is in such short supply that, even at £21,045, it represents a solid investment. It's a car of subtle experiences and understated good taste (interior notwithstanding). What it isn't is an exciting package at a knock-down price.

Enter the M535i. Subtlety isn't this car's middle name and, in the wet, you wouldn't want to know it. But it is a seriously fast car with seats for four big people and their luggage and secure, entertaining handling in the dry. Its engine is superb, its ergonomics faultless and its driver-appeal sky high. For less than 20 grand, you can't do any better.

Mercedes – is marginally the more powerful on ram pressure alone. No quibbles, though: these are two of the best systems in the business.

NOISE

Our objective measurements say the Mercedes is the noisier car and subjective impressions would certainly bear that out. The Mercedes' 16-valver never gets loud, but its slightly gruff and "metallic" note has a curiously penetrating quality that cuts through the generally low levels of wind and road noise.

If there is such a thing as a mellifluous engine note, the BMW has it: smooth, muted, no hint of

Capacity, cc	3430
Power bhp/rpm	218/5200
Torque lb ft/rpm	224/4000
Max speed, mph	139.1
0-60 mph, sec	6.9
30-50 mph in 4th, sec	7.6
mph/1000 rpm	29.0
Group test mpg	20.4
Touring mpg	24.5
Weight kg	1391
Drag coefficient Cd	–
Boot capacity m³	0.37

FRONT SUS	Independent by MacPherson struts, coil springs, double pivot linkage, anti-roll bar
REAR SUS	Independent by semi-trailing arms coil springs, braking/lift-off compensator, anti-roll bar
STEERING	Recirculating ball, power assisted
BRAKES	Vent discs/discs
WHEELS	Alloy 165 TR 390 mm dia
TYRES	220/55 VR 390 TRX

Capacity, cc	2299
Power bhp/rpm	185/6200
Torque lb ft/rpm	173/4500
Max speed, mph	137.1
0-60 mph, sec	7.4
30-50 mph in 4th, sec	7.8
mph/1000 rpm	22.5
Group test mpg	23.5
Touring mpg	32.5
Weight kg	1258
Drag coefficient Cd	0.32
Boot capacity m³	0.31

FRONT SUS	Independent by MacPherson struts, lower wishbones, coil springs gas-filled dampers, anti-roll bar, anti-dive geometry
REAR SUS	Independent by Mercedes multi-link system, coil springs, gas-filled dampers, anti-roll bar, hydropneumatic self-levelling
STEERING	Recirculating ball, power assisted
BRAKES	Vent discs/discs
WHEELS	Alloy, 7 in dia
TYRES	205/55 VR 15

COMPARISONS

PERFORMANCE

	BMW	Mercedes
Max speed mph	139.1	137.1
Max in 4th	–	125
3rd	101	89
2nd	64	62
1st	37	38
0-60 mph secs	6.9	7.4
30-50 mph in 4th, secs	7.6	7.8
50-70 mph in top, secs	10.7	9.7
Weight, Kg	1391	1258
Turning circle, m/ Turns lock to lock	9.9/ 3.1	10.1/ 3.3
Boot, m³**	0.37	0.31

**as measured by Motor

COSTS AND SERVICE

	BMW	Mercedes
Price, inc VAT & Tax, £	19,495	21,045
Insurance Group	8	8
Group test mpg	20.4	23.5
Touring mpg	24.5	32.5
Fuel grade (stars)	4	4
Tank capacity, litres	70	70
Major service int, miles	†	12,000
Intermediate, miles	†	6,000
No of dealers	148	106
Set brake pads (front)£*	46.62	34.90
Complete clutch £*	189.76**	136.45**
Complete exhaust £*	555.61	298.94
Front wing panel £*	141.85	96.14
Oil filter £*	4.43	3.62
Starter motor £*	116.55**	237.01
Windscreen £*	140.36	94.53
Service time, hrs change clutch	2.75	3.9
change water pump	1.08	2.3
major service	4.25	2.7

*inc VAT but not labour costs
†service indicator
**exchange

STANDARD EQUIPMENT

	BMW	Mercedes
Seat back map pockets		●
Courtesy light delay	●	●
Boot light	●	●
Central door locking	●	●
Remote fuel flap release	●	
Electric mirror adjust	●	●
Heated mirrors	●	
Intermittent wipe (variable)	●	●
Programmed wash/wipe		●
Headlamp wash/wipe	●	●
Driving lamps	●	
Fog lamps	●	●
Illuminated vanity mirror		●
Seat height adjustment (driver)	●	●
Seat tilt adjustment (driver)	●	●
Adjustable upper seatbelt mounting	●	●
Adjustable steering column	●	
Rear seat head restraints		
Rear centre armrest	●	●
Rear seat belts	●	●
Rear comp heating		●
Tinted glass	●	●
Power assisted steering	●	●
Self levelling suspension		●

RATING

	BMW	Mercedes
Performance	●●●●●	●●●●
Economy	●●●	●●●●
Transmission	●●●●	●●●
Handling	●●●●	●●●●●
Brakes	●●●●	●●●●●
Ride comfort	●●●●	●●●●●
Accommodation	●●●●	●●
Boot/storage	●●●●	●●●
At the wheel	●●●●●	●●●●
Visibility	●●●●	●●●●
Instruments	●●●●	●●●●
Heating	●●●●●	●●●●●
Ventilation	●●●●	●●●●
Noise	●●●●	●●●
Finish	●●●●	●●●●●
Equipment	●●●●	●●●●

Excellent ●●●● Good ●●●
Average ●●● Poor ●● Bad ●

FIVE Ms JUSTIFIED

Behind the red, purple and blue colours of BMW's Motorsport lies a special breed of sporting car. John Picton finds out how an M-car earns its stripes

The M-cars. A mean and magnificent collection; 1300 horsepower and £200,000 worth of BMW's finest auto engineering. Power, performance, prestige and — inevitably — price mark the five models to wear the red, purple and blue striped logo, but it's still a disparate collection, from mildly tweaked road car to thinly disguised racer.

Yet there are some common threads that run through from that first M-car, the stunning M1, to the latest, the exhilarating M3. They aren't as thickly woven as those between, say, Ford's XR-series models or Audi's various quattros, but they still pick out a delicate pattern of ideas and principles that could be summed up as the M-car philosophy.

It would be easy to think that motor sport is the real link between the cars. But that's too simple; BMW was racing long before the M-range began and cars like the 2002 Turbo or 3.0CSL would be as deserving of a motor sport M of approval as any of the existing models.

No, Motorsport with a capital M — and with it, eventually, the M-cars — came to BMW in the '70s with Jochen Neerpasch, who arrived from Ford to set up BMW Motorsport GmbH. The GmbH is important; it's the equivalent of limited company status and stakes out Motorsport's position as a separate company with considerable independence from its watchful parent.

Motorsport GmbH was a recognition by BMW that competition was becoming big business, both technically and commercially. If the investment was huge and growing ever larger, so too were the rewards. Not only could the new company improve sales and corporate image, but it could add new inputs, from swish leather jackets to improving the production cars.

Taking this idea to its logical limits, what better than for Motorsport to develop its own model variants? They could be used for homologating special competition parts while offer-ing the sort of uprated performance that would appeal to BMW's keener customers. An appealing plan but one whose first stage, the M1 project, was its biggest, boldest and, ultimately, most embarrassing. The M1 chapter of accidents was borne of BMW's long-nurtured desire to defeat its great track and road rival. At the time Porsche's dominance was crushing the life out of sports car racing. However, the talk in the '70s was of a new silhouette formula, or Group 5, that would stimulate manufacturer interest and provide effective new opposition.

Loosely speaking, the silhouette race cars would have to retain the profile and mechanical configuration of the minimum 400-off Group 4 production cars from which they came; otherwise, pretty much anything was allowed. Neerpasch knew that a successful silhouette car would have to be based on a mid-engined production car. BMW had already toyed with such a car, showing off a Paul Bracq-designed mid-engined styling exercise in '72. Bob Lutz, then BMW's vice president of sales and a keen motor sport enthusiast, saw the value in a project which could put one over on Porsche, give the company a small number of profitable exotic coupés to sell and, above all, create for it a prestigious new image. But Lutz made one big mistake: he sub-contracted the job to Lamborghini. The Italian company offered both the production capacity and the experience of small-scale manufacture.

The plan was for Giugiaro's Ital-Design, which had styled the car, to produce the glass-fibre body, for Munich to provide the mechanicals and for Lamborghini to manufacture

BMW's approach *to supercar production has changed since the days of the M1. The M3 leading it (right) is clearly a 3-Series despite its being a homologation special. Its engine (below) is a twin-cam four developing 200bhp*

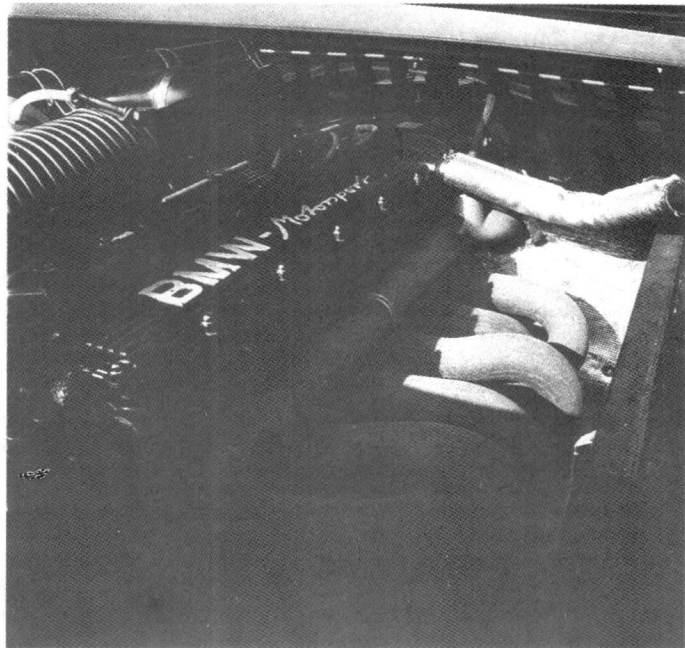

The M1's twin-cam *six was longitudinally mid mounted and powered the car to 162mph*

most of the chassis frame and complete assembly. The schedule called for a public debut in the 1978 Geneva show, a competition debut at Le Mans three months later and a total of 800 cars to sell. But Lamborghini was in financial trouble, the schedule slipped badly, BMW dropped the Italian company and Baur and Motorsport took over the work. It was still a homologation disaster and not finally passed until 1981.

Lutz, the M1's great supporter, had long gone to Ford and Neerpasch went in '79 but not before, in the face of rapidly diminishing management enthusiasm, he had cleverly resurrected the M1's promotional appeal with Procar — two seasons of the world's fastest-ever one-make race series when GP stars and top private teams battled out a BMW entree to each F1 round. It was good stuff but it was not the same as beating Porsche and selling 800 supercars. When BMW finally pulled the plug, just 456 M1s had been built and their competition successes were modest.

Ironically, the M1 has achieved its greatest fame since its demise. Its scarcity makes it one of the great collectors' cars. M1 owner Tim Hignett, who collects BMWs as well

as selling them through his L&C dealership at Tunbridge Wells, Kent, values his immaculate machine, standard down to the Becker radio, at £80,000. He's not boasting.

Though the M1 is different from any BMW before or since, there is much about it in the BMW tradition. The quality, to begin with. You could forgive some shortcomings in a car with competition roots and troubled origins, but Tim Hignett's car is flawless and strikingly free of the rattles and creaks you almost expect to hear in a mid-engined exotic.

It's a cliche to call the M1 an Italian supercar with German attention to detail and yet that is just what it appears to be; immaculate paint, ripple-free grp body (bonded, not bolted, to the chassis), neatly executed details. Only inside the carefully hand-finished leather interior does a set of one-off dials and switchgear provide a striking contrast to BMW's familiar ergonomics. Even here, the constant low buzz of the airconditioning fan is really the only obvious touch of Latin laxity. Inside you discover, as well, a comfortable driving position with few of the usual mid-engined-imposed compromises, save slightly offset pedals and less than ideal rear vision. You're conscious though of the M1's size. By mid-engined standards this is a big beast: 172ins long, 72ins wide and ▶

THE M CARS' PERFORMANCE

	M1	M3	M5	M535i	M635CSi
Top speed (mean)	162mph	139mph	147mph	141mph	156mph
0-60mph	5.5sec	6.1sec	6.0sec	7.4sec	6.1sec
0-100mph	13.0sec	19.0sec	15.4sec	19.5sec	15.6sec
0-120mph	20.2sec	29.8sec	23.3sec	30.8sec	23.4sec
Standing	13.6sec	15.7sec	15.4sec	15.6sec	14.7sec
¼ mile	103mph	92mph	100mph	89mph	97mph
30-50mph in top	7.8sec	9.8sec	9sec	11.1sec	9.6sec
30-50mph in 4th	6.3sec	6.8sec	6.1sec	8.3sec	6.6sec
50-70mph in top	8.4sec	9.4sec	8.3sec	12.3sec	8.8sec
50-70mph in 4th	6.5sec	6.4sec	6.3sec	8.0sec	6.4sec
70-90mph in top	9.6sec	9.6sec	9.7sec	13.3sec	10.1sec
70-90mph in 4th	6.3sec	6.5sec	5.9sec	8.4sec	6.0sec
Overall consumption	17mpg	20.3mpg	16mpg	17.7mpg	17mpg

◀ on a 102ins wheelbase.

The M1's engine is pure BMW. A 3453cc (93.4 × 84.0mm) production development of the 24-valve six that had powered the Group 2 CSL, the straight six is dry sumped to reduce height and it sits vertically and in line behind the cockpit. The Group 4 race engine produced 470bhp at 9000rpm, while the ultimate goal was a twin-turbo Group 5 version with at least 800 horsepower. Productionised, with chain rather than gear-driven twin cams and timed injection, the road version produces 277bhp at 6500rpm. BMW claimed 162mph with 0-60mph in 5.4secs.

Like all BMW sixes, the 24-valve has more than just power. It's supremely tractable, pulling steadily from little more than 1000rpm with an easy-natured flexibility that belies its racing pedigree. The real surge comes once the tachometer needle has moved into the second half of its climb around the clock face. Then the M1's push forward is mighty, chain drives jangling behind you, exhaust howling.

The M1's purpose-determined spaceframe chassis, with wishbones and coils all round, ventilated discs and rack-and-pinion steering, leaves little need to question its handling and roadholding; they're good. The steering, with a tendency to weave as the front wheels follow surface changes, needs some familiarisation but once understood gives taut control. A little less nimble than some of the smaller, shorter mid-engined coupés, the M1 has an accurate, measured cornering style; on fat 205/55VR16 front and 225/50 rear Pirellis the grip is glue-like. Ease the throttle through a right-left flick and the tail will drift out a little, but you'd need to be a brave owner or a Procar racer to drift this big car on the throttle on a regular basis.

The ease of driving almost disguises its massive performance. The engine is as docile as a family Ford when need be, the ZF 'box, despite an awkward dog-leg first, has a shift quality the match of most mid-engined cars, the pedals are light, noise levels are not excessive and the ride on the big P7s is remarkably compliant.

Unlikely as it looks, the little M3 is the M1's direct heir; an unabashed homologation special, created for motor sport success first and foremost and powered by a new competition-derived four-cylinder engine. So the M3 takes over the M1's mantle as the focus of the company's own racing efforts. Like the M1, the M3 was developed by BMW Motorsport but the needed 5000 a year production volume meant the parent company could build the car without the need for questionable outside assistance.

Though it looks the boy racer, with its quattro-like pumped out arches and rear deck spoiler, the M3 has all the sophistication and refinement of its less well endowed cousins. It's still a surprise to find a competition-orientated car so cultured. A surprise to find that the bulged arches are all-new steel panels; to see the way the front and rear air dams have been

M535i: *runt of the litter but still impressive*

M5: *transformed from the disappointing to the sublime*

M635CSi: *M6 tag being saved for an even faster car?*

designed to incorporate a new bumper assembly (legal worldwide); to note how the rear screen angle has been altered and the rear deck line raised with plastic panels to improve aerodynamics. Attention to detail goes as far as bonding in the screens to improve shell rigidity.

But the real surprise comes in the driving. Not in the performance of the 200bhp 16-valve 2.3-litre engine but in the everyday docility and all-round refinement of the Motronic-controlled unit. It isn't as silky as the six but for a race-ready big four it's impressive. And, like the M1, the refinement and tractability cloak the performance.

The M3 is free, too, from circuit-induced compromises; the ride is taut but only feels stiff around town, while the interior is that of the 3-Series at its most luxurious, including sports front buckets and leather seat trim.

The M3's most impressive aspect, though, is the transformation to handling and roadholding. What looks on paper to be tuning rather than changing the suspension has produced a 3-Series that steers tautly, rolls little and, above all, handles neutrally. Uprated brakes have also made a substantial improvement to stopping. This car is a treat, and a revelation, to drive.

Just the same is true of the M5, another car that the Motorsport touch has transformed from the disappointing to the sublime. The fast four-door is currently the only road car that Motorsport both developed and builds. The M5 comes down a separate line at a rate of 250 a year.

This M machine lacks the competi-

tion connections — though it is raced in some markets — and is fundamentally a road car, albeit one which Motorsport has changed considerably. The company starts with the engine, which is a modernised, wet sump version of the M1's unit. It now has slightly more power (286bhp) and torque, but the real change is the adoption of Bosch Motronic injection and electronic management.

Smooth and stunningly powerful, the multi-valve six has proved a boon to the M-cars; from the original M1, it's since been slotted into both the M5 and the M635CSi to turn both into the Porsche-eaters BMW badly needs. The pair have similar performance, but it's the saloon that's the more exciting, with a tautness and immediacy of response that the bigger coupé can't match.

You won't find much on paper to explain the superior handling balance of both cars over their mainstream counterparts; revised front-end geometry, thicker anti-roll bars, gas dampers, stiffer springs, substantially fatter tyres (P700s on the M5, TRX on the M635), a limited slip diff and — in the M5 — putting the battery into the boot to improve weight distribution. If you think suspension sorting of this order is a simple business you've only to drive the M535i.

This is the runt of the M-series litter. It's a second generation combination of 5-Series body and the 12-valve 3430cc six of the 6 and 7-Series cars. The result is a much more favourable balance of power to weight, even though its 218bhp seems mild after the M5's 286bhp. With fat 220/55 Michelins plus a limited slip diff it has considerably more roadholding than standard. Yet Motorsport's involvement here went little further than the garish M-Technic spoilers.

And you can tell when you drive the hot 5s together. The M5 is sharper, has less roll and a near neutral balance; the 535 feels taller, more vague and uncertain as it nears its oversteering limits.

Once you remove this car from the M-series equation, the links between the M-cars become more apparent. And if you call the coupé the M6 instead of by its clumsy M635CSi title, you have M1, M3, M5 and M6 — a line that starts to look much more logical.

But the question remains; what is a BMW M-car? The cynic would say the improved roadholding and more neutral handling ought to apply to every BMW. That point made — and Motorsport argues that its improvements do gradually rub off on the high volume models — the M-cars do have features in common, things like build quality, refinement, smoothness and comfort.

But to end with a couple of questions; why is the M635 not called the M6? Perhaps because a quicker, sportier version is on the way? And will the line culminate with an M7 version of the new 7-Series? With, perhaps a Motorsport version of the coming V12? *That* should be an M to remember . . .

Will the line culminate with an M7 version of the new 7-Series? With perhaps a Motorsport version of the coming V12? That would be an M to remember

The ultimate Q-Car in the BMW range, the M5, and a look at full fresh air motoring from the same stable

BMW M5

I SUPPOSE it's indicative of the performance of the new M5 that BMW chose to launch it on German autobahns where there are no speed limits.

Fitted as it is with the 286bhp 24-valve 3,453cc straight six engine first seen in the M635CSi, one might expect it to be quick, but BMW claim it as the fastest four-door saloon currently available in Britain with its 153mph top speed and 0-60mph time in 6.2 seconds claimed performance figures.

Unlike the M635CSi, the restrained lines of the 5-series saloon give this car a completely different character and the intentional lack of any ostentatious body addenda perhaps makes it the ultimate 'Q-car', a true wolf in sheep's clothing. In fact, the only external tell-tales that might show the M5 to be any different from its £9,990 518i brother are the 7.5J×16 BBS alloy wheels and Pirelli P700 tyres and a deeper front spoiler.

We were able to drive the new car on *autobahn* and across country and in both environments it performed well. At 70mph the car feels extraordinarily unstressed; 100mph is equally effortless

Above, new from BMW is the full convertible which sells for £17,840. Full driving impressions coming soon.

Left, understatement is the key to the M5 concept. Its performance is stunning, however

and at these speeds noise levels are very low indeed, save for the muted hum of the engine which I cannot imagine could be anything but music to any driver's ears. We saw an indicated 145mph or so on a clear stretch of road so it seems that BMW's claimed top speed figure might, if anything, be a little conservative, bearing in mind the M635 is good for a little short of 160mph, according to BMW.

With the engine's tremendous power and torque, the superb roadholding, traction and brakes, cross-country driving on A and B roads is very rewarding, and the ABS, fitted as standard, is a reassuring feature in case of emergency.

With the BMW M5 comes a

list of standard equipment that is probably even longer than one's arm; apart from the alloy wheels, P700 tyres and ABS already mentioned, the car comes with air conditioning, electrically adjustable sports front seats, electric sun roof, electric windows, power steering, sports suspension, onboard computer, central locking, a sophisticated Blaupunkt stereo/tape player with electric aerial, electrically heated and adjustable door mirrors, headlamp wash/wipe, velour upholstery to the interior and boot and a built in alarm system to keep it all one's own. A nice touch is the fitment of an elasticated net in the boot to keep luggage from sliding about.

Not surprisingly all this

brings the price of the M5 up to a whacking £32,390, a clear £11,000 more than the 'normal' M535i but still £6,600 cheaper than the M635CSi. By contrast, the top of the 7-series range, the 735i SE, costs £31,750. It is a sobering thought that for this sort of money, one could have a Mercedes 300E *and* a BMW 320i — and still have over £2,000 change!

Quite definitely the conservative appearance of the M5 will make it a winner with company fleet managers, undoubtedly the market area at which it is aimed, which is probably the reason why all 70 of this year's allocation have long since been sold. The waiting list, I believe, is growing fast . . .
Paul Clark

As an out-and-out race car the BMW M3 has already proved itself. The roadgoing version of this 'homologation special' is just as successful.

Bavarian battle-cruiser

Discreet, yet valiant, that's the BMW M3. Timid in town (as required), overtly rampant on the open road (if necessary), and ultra competitive when unbridled for out-and-out competition use, the Bayerische Motoren Werke's latest sporting carriage fulfills its designers' every criteria with pace, pizazz and panache. Unashamedly a 'Homologation special', the M3 has immediately proven itself to be a formidable weapon on the race circuits, and put the Munich marque to the forefront of World rallying.

Remarkably, the M3's sporting aspirations do not detract from its practicality as a road car of rare character but, once behind the wheel, do not expect discretion to get the better part of valour. Well, not too often anyway...

I say remarkably for this is the number one son of the legendary BMW Motorsport GmbH concern, headed these days by the redoubtable Wolfgang-Peter Flohr. His team of crack engineers effectively evolved the M3 backwards, designing a Group A racing car, then productionising it rather than merely adapting an existing roadster. Demand for their little gem was inevitably going to be high from day one, even outside the overtly image-conscious German market, thus the pre-requisite 5000 examples were completed in Munich for homologation papers to be rubber stamped on Mar 01, this year.

While BMW Motorsport's first series production project (the rakish M1 sportscar – of which 456 examples were built from 1978-'81, and best remembered in its fabulous 'Procar' racing trim – could never have sustained anything more than specialist clientele) bears more than a passing resemblance to its parent company's long-established 3-series models (a 16 car line-up ranging from the 'humble' 316 at £8995 to the hairy £19,895 325i Convertible), the M3 is subtly different from stem to stern. And, of course, the restyling work is not the manifestation of aesthetic whim. Every detail of

this 'race-roader' – with due deference to Dr Spooner road-racer does not reflect the rationale behind it – answers an important criterion in the anatomy of a Group A winner.

Apart from the deep, oil cooler ducted, air dam and neatly integrated swathed wheel arches and sills, the M3 boasts a distinctive tail spoiler atop its slightly raised and shortened boot lid. All of these features – one hesitates to call them changes – are understated, most typically of the revered marque, whose former brashness was exhausted on the overtly bewinged 'Batmobile' CSLs of the '70s. The most interesting and immediately noticeable realignment of the traditional 3-

series silhouette has been effected by raking the rear screen to further enhance the aerodynamic properties of the shell. The resultant drag coefficient of the package is quoted at 0.33.

Beneath the bonnet nestles, no lurks, a delightfully free-revving short-stroke 2.3-litre 16-valve evolution of the BMW's trusty M12 four-cylinder power unit. Developed through extensive and ultra-successful racing experience with 2-litre F2 single-seaters, as well as touring cars in many guises, the compact engine punches out a meaty 200bhp in M3 road cars, and some 50 per cent more in GpA race or rally guise. Reliability and longevity are keynotes of this musical four, the

The 2.3-litre engine is a derivative of the winning 2.0-litre F2 power plant, and turns out a healthy 200bhp in road trim.

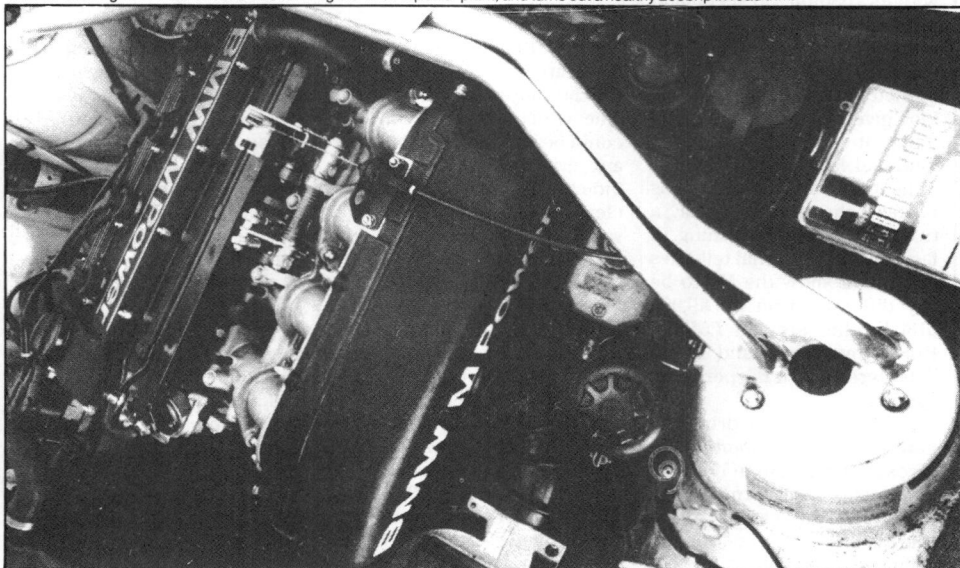

quality of whose build is outstanding. The engine is fed by Bosch's ML Motronic fuel injection.

Suspension is taut, competition springing endowing the M3 with a strikingly squat, aggressive, stance. Custom-rated gas dampers and sturdy anti-roll bars at each end ensure that the exemplary roadholding ability, and agility, are optimised. To balance out goodly measures of both power and handling, huge ventilated brake discs all but fill the road car's 15×7in wheels (the flared arches accommodate 10in wide rims for racing purposes), and have ABS anti-locking assistance. Hydraulic power steering, speed sensitive, is perfectly weighted, providing plenty of feedback to the large 15in steering wheel.

The M3 is only available in left-hand-drive form, which may dull its appeal to some, but anyone familiar with single-seater racing cars will feel instantly at home in its functional all-black cockpit, the chunky gear lever naturally falling to hand. Getrag's proven 5-speed close ratio box is specified, which means that first is on a dog-leg, hard left and back, as in 5-speed Hewland transaxles. It may be a touch notchy at first acquaintance, but quickly becomes second nature to newcomers to the pattern. If, perchance, you hook second gear by mistake, the engine has quite sufficient torque not to leave you floundering at the traffic lights.

But it is for the green 'go' light that the machine was envisaged. With engine revs correctly matched to surface, and a corresponding ration of right foot, the M3 surges forward to 60mph in a shade over 7s, its 25 per cent limited slip diff optimising the grip of Pirelli's chunky P600 rubber and catapulting the car straight as a die once the power has been taken up. Less patient pilots can break traction at will, to the ultimate detriment of acceleration times.

A little more restraint can bring electrifying performance however, and conserve tyres, not that these – like 20mpg fule economy – can be a major consideration to those spending £23,550 for the pleasure of M3 motoring. The rewards of driving the car quickly and smoothly are exceptionally high, for its inherent balance, tremendous mid-range flexibility, lightning quick response to turn-in commands and solid, reassuring, feel inspire confidence beyond the realms

of normal everyday motoring. The M3 is a competition car, and does not let you forget it.

As with anything this potent, it is critical not to become over confident at the helm. The sprinter's speeds can be deceptive – particularly on fast, sweeping, cross-country routes – and its in-bred characteristics will certainly flatter the ego of the average driver, if not his driving ability. The M3's limits

to be known, before the advent of 'high tech' nomenclature as bucket seats. Bonded glass keeps wind noise to a minimum, while the intrusion of road roar is well damped and unobtrusive. Fastidious quality control at the factory has left the car free from irritating rattles, as one should expect at this, or any, price when buying new.

The driving position allows effortless coverage of long distances,

Porsche on that front (quite), or the carrying capacity of the bulkier Merc or Sierra, it is arguably the most interesting option of the foursome to drive and, for the moment at least, the most exclusive on British roads. All four machines have similar top speeds, incidentally, in the 140+mph bracket.

It is abundantly clear, from the moment one climbs aboard, that the M3 is a magnificent drivers' car, and

Lined up alongside BMW's tame-looking M5, the M3 is more instantly recognisable as a sporty – even race – car.

are indeed extraordinarily high, but must be courted with respect, and over-stepped with caution on race track or rally stage only. These arenas are, after all, this BMW's natural habitat. Not everyone, thankfully, is a Roberto Ravaglia, Winni Vogt or Bernard Beguin – just three of BMW's trump cards in this season's World and European Touring Car Championships, and World Rally Championship.

For a racing car, the M3 is surprisingly civilised as a road burner. Noise levels are comfortable at all speeds, and front seat occupants at least are cosseted in anatomically supportive leather upholstered luxury – what used

whether at 'press on' or more leisurely pace. Visibility is good, and the competently simple, if dated, dash clearly legible. A large clutch rest in the spacious footwell is welcome.

The BMW will undoubtedly be compared with the fire-breathing Ford Sierra RS Cosworth (much cheaper at £17,412) and Mercedes-Benz's more stylish than sporty 190E 2.3-16 (£24.670), not to mention the less practical but super suave Porsche 944S (£25,303), and in my opinion it scores well in this exalted company. While the M3 cannot match the turbocharged 2-litre Ford on initial acceleration, or the sleek two-plus-one-ish

with BMW's reputation, its outstanding performance will last. Where racing improves the breed, and the breed is geared to competition, then manufacturers following BMW's lead can expect their future road cars to develop apace. If the M3 can withstand the most rigorous punishment in both race and rally circles, you can safely assume that the same hardware and engineering will not wilt in day-to-day use. And that's just part of the message that BMW is trying to put across through its ever innovative approach to motor sport, in which the M3 is the latest in a long line of winners – road, race and rally! ∎

The latest in a long line of winners for the German company, is shaping up well on both track *and* field.

BMW M3
£23,550

SPECIFICATION

Cylinders/capacity	4-in-line, 2302cc
Bore/Stroke	93.4/84.0mm
Valve gear	Twin ohc, 16 valve
Fuel system	Bosch ML Motronic fuel injection
Power/rpm	200bhp (DIN) at 6750rpm
Torque/rpm	177lb ft at 4750rpm
Gear ratios	1.00, 1.26, 1.77, 2.40, 3.72:1
Final drive	3:25:1
Steering	Rack and pinion, power assisted
Brakes	Discs (ventilated in front) with ABS
Wheels	Alloy, 7J x 15ins
Tyres	205/55 VR15
Suspension (F)	Independent by MacPherson struts, coil springs and anti-roll bar
Suspension (R)	Independent on semi-trailing arms, coil springs and anti-roll bar

DIMENSIONS

Length	171.1ins
Wheelbase	100.9ins
Track (F/R)	56.0/56.4ins
Width	66.1ins
Weight	24.6cwt

PERFORMANCE (*Autocar* figures)

Maximum	140mph
0-60mph	7.1s
0-100	19.0s
50-70mph (4th/5th)	6.4/9.4
Fuel consumption (urban/56mph/75mph)	18.3/22.3/26.4mpg
Test consumption	20.3mpg

BMW M6

America, rejoice! The meaty, beaty, big, and bouncy M6
is here, and it's yours for $56K.

BY JOHN PHILLIPS III

Ann Arbor—Drive around Southern California—indeed, anywhere there's a heavy density of sports cars—and you'll see BMWs with "M" badges. Lots of them. Most of those cars became M-machines by mail. They're as much BMW Motorsport's progeny as Candice Bergen is a singer. Now, finally, among the swarm of falsely badged M-cars in America, there are a few that really are, if you'll pardon the pun, *M-azing.* The M6, the M5, and the M3 have arrived, ticketed with base prices that reflect their limited numbers: $55,950, $43,500, and $35,000.

We were among the fortunate few to test an M6 within weeks of its arrival on these shores; it was the first of 1200 to be sold here. For us, the timing couldn't have been better. We were concluding our Four Seasons test of the 635CSi. With both 6-series cars at our fingertips, back-to-back comparisons were a snap.

The M6 (and the M5, for that matter) is blessed with the fast-revving S38 3.5-liter engine. Even if you aren't considering a BMW, you owe it to yourself to eyeball this engine in the flesh (in the alloy?). It ought to

PHOTOGRAPHY BY KEN OSBURN

be in a museum for modern art. At a Sunoco station in Bowling Green, Ohio, the attendant, who originally planned to check the oil, got an eyeful and practically lost his mind: "Keeee-*rist!* Jim, come out here and *look* at this." Jim arrived on the trot, followed by Bob, Roy, Maynard, and Billy Joe. Oil was never checked. The scene was repeated at each fill-up.

The object of all the envy is a DOHC 24-valve in-line six that produces 256 bhp. That's 75 bhp and 29 pounds-feet more than the 635CSi can muster. The M6's engine, which lost 30 bhp during its federali-

M6 includes special front spoiler and deck lid lip, and sits a half-inch closer to terra firma than regular 6-series BMWs.

BMW M6

zation, has a shorter stroke and a slightly larger bore than the 635CSi's familiar 3.4-liter SOHC six. The M-car's valves are in a "V" formation, actuated by cup-type lifters above the hemispherical combustion chambers. The compression ratio, at 9.8:1, is sky-high by today's standards; 91-octane premium unleaded, as you might expect, is the mandatory diet. Caught short in the heart of Ohio farmland, we were forced to give the M6 a drink of 88-octane low-cal. We're relieved to report that the engine uttered not a solitary ping.

In fact, the M6 makes only musical noises, thanks to headers, a low-restriction catalytic converter, and dual exhausts. The symphony of sounds emanating from the in-line six's individual throttles is as hair-raising as anything this side of a Cosworth DFV. The M6 encourages participation. Says the wife of one of our test drivers: "When he drives the car, he makes mouth noises that mimic the engine. It's eerie."

It seems funny now, but we approached our first drive in the M6 with some caution. Since the car's introduction at the Frankfurt show in 1983, European magazines have led us to believe that simply inserting the key in the ignition would cause the M6 to erupt in blue tire smoke, bellowing and slewing sideways out of its parking slot.

Of course, we needn't have worried. The engine is docile and eminently civilized, perfectly content to chug around town in the first three gears. It's so tractable, in fact, that only premeditated mechanical abuse would prompt the rear tires to spin on dry pavement. One's first tour in an M6 is like spotting Muhammad Ali in an airport. Your initial reaction is, "Hey, he's not as big and burly as I expected."

The M6's forte isn't evident until you pick a gear and *leave* it there. Second gear, for example, will happily carry you cleanly to 60 mph. Mash the accelerator at any engine speed above 2500 rpm, and the power is delivered in an insanely smooth arc that extends right up to the redline. It's uncanny. We can't think of any other engine that has a nearly flat 4000-rpm power band. Who needs turbos when a four-valve in-line six is capable of this kind of power and tractability? (Yes, Bavarian engineers have been asking that question for several years.)

BMW insists that the 256-bhp M6 can tackle 60 mph in 6.8 seconds and, barring a strong head wind, will achieve 150 mph. We consistently recorded times in the low sevens. Still, that's 1.2 seconds faster than our 635CSi and the equal of a Ferrari Mondial 3.2. To be perfectly fair, our inky black M6, with only 1500 miles on the clock, was barely past its prescribed break-in period.

Although the 6-series BMWs are renowned for subtle styling, there's no mistaking the M6. Other BMW owners pull alongside and gawk shamelessly, clogging the passing lane and leering during red lights.

Once they're persuaded that the M6 isn't sporting fake badges—a few blips of the throttle seem to remove all doubt—fellow enthusiasts wave frantically and want to talk: "Wow, it's real, isn't it? A gray-market car? How much?" The conversation usually falls deathly silent when the price is revealed.

One of the reasons the M6 is so easy to spot is that, like a cat ready to pounce, it sits a half-inch closer to terra firma. It also comes standard with a larger front spoiler (oil cooler tucked therein), an inch-high rubber deck lid lip, and massive Michelins. "One flat tire, and there goes a week's salary," said our exuberant gas-station attendant in Ohio, and he's probably right.

But don't laugh at the attendants. You'll want to be friendly with all of them within a 100-mile radius. The EPA assigns the M6 a city-driving rating of only 10 mpg, and the environmental guys aren't fooling. During 500 miles of what was, admittedly, spirited motoring, only once did our M6 achieve mileage in the teens. That's shabby enough that you'll develop a spicy new vocabulary to describe the sixteen-gallon tank. Perhaps self-conscious about the problem, BMW removed the fuel economy meter that usually occupies the lower half of the tach. Why be reminded at every light, eh? Of course, after you've paid the $2250 gas guzzler penalty, you won't be inclined to forget anyway.

If you can afford an M6, you can

Sumptuous interior includes glove-soft leather skins, new tachometer, and flawless five-speed.

Rear seat space is compromised by an ungainly bin that houses twin air-conditioning ducts.

BMW's DOHC, 24-valve, alloy-head engine produces 256 bhp and is a work of modern art.

afford to buy a lot of gas. We understand that. The problem is simply this: Averaging 10 mpg, the M6 has a cruising range of only 166 miles.

The standard drivetrain includes the Getrag 280 heavy-duty five-speed transmission, with specific gearing for this engine, and a 3.91:1 limited-slip differential. You can't have an automatic transmission in an M6. Shift action is typical of BMW; it's beyond reproach. The clutch, however, is heavy, and one of our drivers said the pedal was difficult to hold at red lights.

At 100 mph in fifth gear, our M6 was turning 4200 rpm. It cruises at such revs effortlessly, but the driver must be willing to accept trade-offs: considerable noise—a combination of a lot of valves flying hither and thither and a rip-snorting exhaust boom—as well as the aforementioned fuel fiasco. Mind you, it's not often in America that 100-mph cruising is feasible, but it was of interest to us because our long-term 635CSi test car will hum along at 100 mph at 1000 fewer revs, at least when its automatic shift lever is in the fourth-gear "E" position.

Complaints? Yeah, a few, mostly niggling stuff. But for sixty grand, you expect no niggles, right? We were miffed to find no map light, for example. After all, the M6 encourages you to drive down dark mountain roads. It would be nice to know what road you've taken. It's also worth noting that the most often used quadrant of the speedometer—from 50 to 90 mph—is largely obscured by the steering wheel, as are the radio's on/off and seek switches.

Far more bothersome, however, is the interior room—a happy compromise in the 635CSi, but downright skimpy in the M6. The standard electric sunroof and the hand-stitched leather headliner have reduced the headroom. A rear-mounted battery consumes the right quarter of the trunk. (Hey, the engine bay is filled to the gunwales with polished cam covers, intake tubes, and such; no room for batteries.) Even more offensive is an absurd breadbox-sized abomination that has sprouted between the rear seats. It houses two air conditioning ducts and a rack for stereo tapes. Nobody knows why. No longer can you fling a briefcase onto the back seat; it won't fit.

Remember, the 6-series design is now ten years old. BMW publicly confesses that cars with greater interior room are a priority—witness the new 7-series sedans and the spy photos of the 5-series replacements.

The M6 is a long-legged tourer, for sure; a luxo-cruiser with high-speed capabilities, without doubt. But a sports car? Not really. At 3600 pounds, it's too fleshy in the inseam to indulge in close-quarters acrobatics. Yet the handling, for a car of this bulk, deserves no criticism. Directional stability is faultless; the steering, while slightly heavy, is fast and surgically accurate, especially in the first few inches of movement off center; and the aggressive self-centering allows the driver to concentrate on less serious matters, like focusing the rear A/C outlets on something other than his right kidney.

Mild understeer is the watchword, perhaps as a consequence of the firmer progressive-rate rear springs. But it's our guess that the M6's improvement in handling over the 635CSi is largely achieved via wider rubber. (The TRX-sized wheels, at 16.3 by 8.3 inches, are wider, too, which required a slightly recalibrated ABS.) The suspension has simply been meticulously tuned—"optimised," as the Brits put it—to deal gracefully with the fatter contact patch. Cars with tires of this breadth usually hunt and dart, unwilling to track a straight line on bumpy roads. BMW has not only circumvented those faults but has also somehow dialed in a remarkably supple ride, even over Michigan's Godzilla-sized expansion strips. The ride-versus-handling trade-off, here, is our idea of perfection.

Bilstein gas-pressure shocks are fitted all around, with twin-tube versions at the front. Extra steering caster has been added, and the front brake rotors have been enlarged slightly to 11.8 inches. The four-wheel discs do such a smooth and competent job of arresting the car's considerable mass that we never once triggered the ABS on dry pavement. You'd have to be in an eye-bugged panic mode to do that. We did notice, however, that the first half-inch of brake-pedal travel tends to *grab* the rotors, rather than gingerly clutch them. The effect is particularly noticeable when, for example, the brakes are delicately tapped to disengage the cruise control. An annoying lurch ensues, and it's always enough to startle passengers.

What else? Well, there's not time enough for us to enumerate the luxurious fitments—things like hand-stitched Nappa leather, a snazzy tool kit in the trunk, heated front seats, mirrorlike metallic paint, and halogen foglights that actually penetrate fog. Simply write your check for $58,750 (not counting tax, destination, title, or insurance), and you really don't have to waste time studying "option groups." What you *might* want to study, however, are the cars in the M6's price class: the Cadillac Allanté, the Ferrari Mondial, the Jaguar XJ6 Vanden Plas, the Porsche 928, and the Mercedes 560-series. A heady bunch. The M6 equals any of those cars in luxury; it is surpassed by only the 928 and perhaps the Ferrari in sheer performance. That makes it one of the best cars in the world.

In these days of mechanical one-upmanship among car companies, the term "badge engineering" has come to represent something odious to most enthusiasts. It's nice to know that badge engineering at BMW stands for truth, justice, and the Bavarian way. Buy an M6 and you have our permission to sneer cruelly at the clods with the fake badges. ⬤

BMW M6
Base price/price as tested $55,950/$58,750

GENERAL:
Front-engine, rear-wheel-drive coupe
4-passenger, 2-door steel body

POWERTRAIN:
24-valve DOHC 6-in-line, 211 cu in (3453cc)
Power SAE net 256 bhp @ 6500 rpm
5-speed manual transmission

CHASSIS:
Independent front and rear suspension
Power-assisted rack-and-pinion steering
11.8-in vented front, 11.2-in vented rear disc brakes
240/45VR-415 Michelin TRX tires

MEASUREMENTS:
Wheelbase 103.3 in
Curb weight 3570 lb
Fuel capacity 16.6 gal

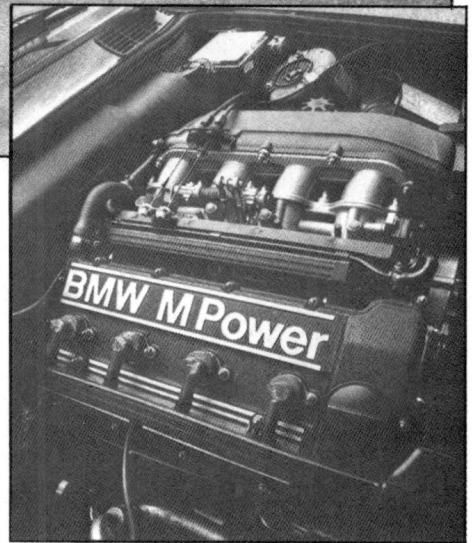

BMW M3

Don't ever think "yuppie" again.

• Let's put an end to the BMW-yuppie link. These days, all you have to do is whisper "BMW," and everyone immediately thinks "yuppie." Enough of the yups! That overpublicized group of consumers, who lust after Bimmers as they do any object perceived to confer status on its owners, could never fully appreciate the car you see on these pages. We don't mean to say that the young urban professionals won't love the new M3. After all, it's got that famous badge on the hood. And you know, it's a prestigious Eurosedan and everything. But will they realize that the M3 is the latest well-muscled, painstakingly crafted creation from BMW's esteemed Motorsport department? "Nah, but it'll sure look great in front of the condo."

The M3 deserves better. This is not a car for yuppies. This is a car for *us.* In case you haven't noticed, BMW's U.S. lineup has blossomed to include a dazzling array of leather-lined hot rods that beg to be flogged through the twisties and hammered on the superslabs. Gone are the anemic four-cylinder models that nearly ruined BMW's image. Nearly extinct are the Bimmers reserved for social climbers. The Bavarian Motor Works is back on track with a fleet of *drivers'* cars, and the M3 is potent proof of its new direction.

The M3 is the most recent of the broad-shouldered BMW Motorsport models to reach our shores. For those not yet fluent in M-speak, the M-machines are limited-edition, high-performance versions of the 3-, 5-, and 6-series sedans. For several years the M-cars were a treat reserved for European buyers, but since early this year they have been trickling into the hands of hungry American enthusiasts. The M5 and the M6 debuted stateside in February, and the M3 joined the brawny pair in June.

The M3 is available to enthusiasts because of the rules that govern FIA Group A racing. To qualify a car for Group A competition, its manufacturer must build a minimum of 5000 examples of it within twelve months. The rules also strictly limit the modifications that can be made for racing, so most of the performance hardware must be baked into the roadgoing cars. BMW Motorsport clearly knows the recipe for success in Group A road racing: after the fourth of seven European Touring Car Championship events this year, the M3 had already clinched the title.

The M3's racing heritage is immediately apparent in its steroid-injected bodywork. With its aggressive assortment of air dams, body flares, and spoilers, the M3 will quicken the pulse of any boy (or girl)

racer lucky enough to catch a glimpse of one. Most of the new pieces are made of steel, though the rocker panels, the trunk lid, the front air dam, and the rear wing are molded in plastic. The rake of the rear window has been altered for improved aerodynamics, and both it and the windshield are bonded flush with the surrounding bodywork. The net result of all these aero tweaks is a drag coefficient of 0.33, down substantially from the 325i's 0.37 Cd. Perhaps more important, no one will ever mistake the burly M3 for an ordinary 3-series sedan.

The subskin make-over is equally impressive. Like its 3-series siblings, the M3's fully independent suspension has struts in front and semi-trailing arms at the rear, but the coil springs and the gas shocks have been revised and strengthened for race duty. The shorter springs drop the M3 about an inch lower than the 325i. In addition, the front anti-roll bar is attached to the struts rather than to the

control arms, and a beefier anti-roll bar is fitted to the rear.

Formula 1 fans will think they've died and gone to Monaco the first time they lift the M3's hood. Inside sits a normally aspirated, 2.3-liter version of BMW's brutal, turbocharged four-cylinder Grand Prix engine. This is the only remaining four-banger in BMW's U.S. lineup, but it's anything but a prestige-sapping weakling. Hardware enthusiasts have plenty to drool over here: four valves per cylinder, double overhead camshafts, an individual throttle for each cylinder, tuned intake and exhaust plumbing, and a new ML3

Bosch Motronic engine-management system. As further proof that this is no ordinary powerplant, its cam cover and air cleaner are emblazoned with the words "BMW M Power."

Though the "M" stands for "Motorsport," we think "Mucho Power" is more like it. The sixteen-valve four-cylinder turns out 192 hp at a lofty 6750 rpm and 170 pound-feet of torque at 4750. If you think those are remarkable figures for a 2.3-liter, you're right: the M3's ferocious four boasts a higher output per liter than any other normally aspirated piston engine available in America.

What looks impressive on paper feels equally stirring on the road. When its tail is twisted, the 2857-pound M3 dashes to 60 mph in 6.9 seconds and trips the quarter-mile lights in 15.2 seconds at 92 mph. Top speed is an autobahn-tuned 141 mph. That's enough punch to blow off the Mercedes-Benz 190E 2.3–16 and stay neck and neck with the Porsche 944S. Best of all, the M3's power delivery is wonderfully linear; it pulls willingly from its midrange all the way to its sizzling 7250-rpm redline.

Those accustomed to the silky smoothness of BMW's refined in-line sixes, how-

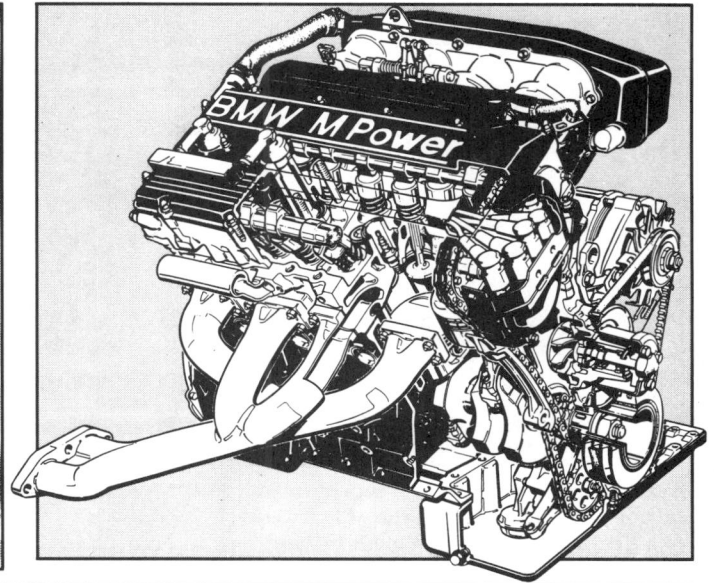

ever, may wince a bit when this engine starts to sing. It is, after all, a highly tuned, relatively large four-cylinder, so a little harshness is part of the bargain. BMW has softened the resonance considerably since we sampled an M3 on the autobahn a year ago, but this engine remains a howler. The noise is a fine, mechanical sound, especially when you're near the very top of the tach, but it's there whether you want it or not.

Most of the time, you won't mind a little kibitzing from the engine compartment, because the M3 is designed for driving with brio. Pushed hard, the M3 comes into its own. The five-speed transmission is tightly geared for maximum go. The chassis is more than a match for the engine, responding swiftly and surely to orders from the helm. Powering through hard corners, the tail stays firmly planted, though there is enough predictable lift-throttle oversteer available to point you back toward your line when understeer begins to be a problem. The standard 205/55VR-15 Pirelli P600 tires don't turn in as crisply as we'd like, but they do stick: the M3 squeals around the skidpad at an impressive 0.81 g.

We had a chance to put in a handful of brisk laps around Connecticut's challenging Lime Rock racetrack, and the M3 proved equal to its breeding. Few road cars can take to the track with such poise. The M3 leaps through the corners like a cat, its feisty engine spinning and spitting until you snatch another gear or the rev limiter grabs it by the tail. Excellent controls help you keep the frenzy in check: the steering is supple and superbly accurate, the shifter has just the right amount of notchiness, and the massive disc brakes—vented in front and equipped with a standard anti-lock system—are always on duty, lap after lap. Our seat-of-the-pants

admiration for the binders was confirmed by our fifth-wheel testing: the M3 clawed to a stop from 70 mph in a mere 179 feet.

The M3 may be a thinly disguised race car, but its creature-comforts list would do most luxury sedans proud. Included are power windows, mirrors, and locks; a power sunroof; air conditioning; a premium AM/FM-stereo/cassette system; a three-spoke, leather-wrapped Motorsport steering wheel; and a nine-function trip computer. Everything is laid out in typically sensible BMW fashion, and the white-on-black analog gauges are among the most legible in the industry. In view of the M3's sporting nature, an oil-temperature gauge has been substituted for the normal 3-series layout's fuel-economy display.

We have mixed feelings about the standard leather seats, however. They offer an adequate range of manual adjustment, and they're dandy for spirited maneuvers. The problem is that they aren't well designed for extended travel. Several staffers complained of a lack of lumbar support, and others suffered from pinched behinds after long drives. We'll give these thrones an overall B. Passengers banished to the rear seats should be either short or masochistic.

All in all, we're smitten by the M3. Our test car was weighed down by a $34,810 price tag—about what you'd pay for a 944S—but the Bavarian beast offers a lot in return. For that princely sum you get a stunningly distinctive design, a generous helping of luxury and quality, and the kind of cool, collected performance available only in German sports sedans.

Enthusiasts who find those attributes tantalizing should get in line immediately: BMW plans to export only 2400 M3s to the United States this year. The supply probably won't be enough to meet the demand, but it will serve to remind enthusiasts that BMW is back in the performance-car business.

Gee, what was that "y"-word again?
—*Arthur St. Antoine*

COUNTERPOINT

• The "batmobiles" are back! With its deep chin spoiler, fabled "bat wing," and forever-revving engine, the M3 reincarnates the spirit of the 3.0CSL racing coupes of the early seventies. This machine is meant for fast, winding secondary roads where its tight chassis can be exercised through generous use of its broad rev band and powerful brakes. Initially, it seems to understeer, but you soon learn that the trick is to brake firmly, then get on the power hard and early. Drive the M3 like that and it behaves in a wonderfully neutral manner, allowing fast, controllable four-wheel drifts that seem as if they could go on forever. As expected, the pedals are arranged perfectly for heel-and-toeing, while the sound of the sixteen-valve four in full cry pumps plenty of adrenaline into your bloodstream.

In small towns, the aerodynamic addenda turn heads while you're just cruising by at 25 mph. And on the open road, necks start snapping as you wail past at over 7000 rpm. The high, stubby, winged tail and the fender blisters give the M3 more than just a passing resemblance to the Audi Quattro Sport. And that isn't bad at all.
—*Nicholas Bissoon-Dath*

I've about had it with you readers. Every time we fawn all over some megadollar exotic, your collective reaction is as predictable as a California sunrise. We're blinded by the smoke, you say. We're in the manufacturer's pocket. We're crazy.

Actually, I'd probably think the same if I were in your moccasins. But here I go again. Stand back, because I'm gonna fawn. Unless you drive the BMW M3, there's no way you're going to understand the synergy at work therein.

Taken separately, the M3's engine, handling, and wild sheetmetal are stellar. Together, their effect is intoxicating. The demonic growl of the engine as it soars to the redline, the racer's-edge moves, and the rally-racer looks push this homologation special into another class entirely.

On paper, the M3 is just an overpriced small sedan with a cramped rear seat. In real life, it's the perfect toy for an upscale closet wild man. The M3 throbs with the soul of BMW's Motorsport division, and I, for one, love it. If you were standing in my moccasins, I'm sure you would, too. —*Rich Ceppos*

Yuppies, keep back! The BMW M3 is keen through and through, and it cries out for only the truly keen to wield its treasury of talents. We car junkies have had it: even if you have sidestepped all the other potential addictions paraded by modern life, do not, under any circumstances, drive this car. It is *so* good that it's hopelessly addictive, and even you, NOx-breath, can't be immune. Nobody who gives a hoot about hardware can resist a tool that does everything well, and a single delirious dose of the M3's dynamics plants an emotional hook that cannot be shaken. It is finesse and muscle and fun packed into one.

The M3 looks right, too, brutish but sophisticated, and the more you drive it, the more you'll love it. In performance, it's right up there with the Porsche 944S (and has a trunk, to boot)! And in feel, stability, handling, punch, and price, it *kills* the Mercedes-Benz 190E 2.3–16. Although I dote on the kick-ass, cocoa-butter smoothness of BMW's 325is, its 27-grand tariff seems a little pricey. And yet another seven Gs seems a mere pittance to pay to OD on the M3.
—*Larry Griffin*

Vehicle type: front-engine, rear-wheel-drive, 4-passenger, 2-door sedan

Price as tested: $34,810

Options on test car: base BMW M3, $34,000; metallic paint, $335; freight, $475

Standard accessories: power steering, windows, locks, and sunroof, A/C, cruise control, rear defroster

Sound system: BMW AM/FM-stereo radio/cassette, 8 speakers

ENGINE
Type . . . . 4-in-line, iron block and aluminum head
Bore x stroke . . . . 3.68 x 3.31 in, 93.4 x 84.0mm
Displacement . . . . 140 cu in, 2302cc
Compression ratio 10.5:1
Engine-control system . . . . Bosch Motronic with port fuel injection
Emissions controls . . . . 3-way catalytic converter, feedback fuel-air-ratio control
Valve gear . . . . chain-driven double overhead cams, 4 valves per cylinder
Power (SAE net) . . . . 192 bhp @ 6750 rpm
Torque (SAE net) . . . . 170 lb-ft @ 4750 rpm

DRIVETRAIN
Transmission 5-speed
Final-drive ratio . . . . 4.10:1, limited slip

Gear	Ratio	Mph/1000 rpm	Max. test speed
I	3.83	4.4	32 mph (7250 rpm)
II	2.20	7.7	56 mph (7250 rpm)
III	1.40	12.1	87 mph (7250 rpm)
IV	1.00	16.9	122 mph (7250 rpm)
V	0.81	20.8	141 mph (6800 rpm)

DIMENSIONS AND CAPACITIES
Wheelbase . . . . 100.9 in
Track, F/R . . . . 55.6/56.4 in
Length . . . . 171.1 in
Width . . . . 66.1 in
Height . . . . 53.9 in

Frontal area . . . . 20.0 sq ft
Ground clearance . . . . 5.0 in
Curb weight . . . . 2857 lb
Weight distribution, F/R 52.9/47.1%
Fuel capacity . . . . 15.3 gal
Oil capacity . . . . 4.6 qt
Water capacity . . . . 11.6 qt

CHASSIS/BODY
Type . . . unit construction with 1 rubber-isolated subframe
Body material . . . . . welded steel stampings and fiberglass-reinforced plastic

INTERIOR
SAE volume, front seat . . . . 45 cu ft
rear seat . . . . 37 cu ft
trunk space . . . . 13 cu ft
Front seats . . . . bucket
Seat adjustments . . . . fore and aft, seatback angle, front height, height, thigh support
General comfort . . . . poor fair **good** excellent
Fore-and-aft support . . . . poor fair **good** excellent
Lateral support . . . . poor fair good **excellent**

SUSPENSION
F: . . . . ind, strut located by a control arm, coil springs, anti-roll bar
R: . . . . ind, semi-trailing arm, coil springs, anti-roll bar

STEERING
Type . . . . rack-and-pinion, power-assisted
Turns lock-to-lock 3.7
Turning circle curb-to-curb . . . . 32.2 ft

BRAKES
F: . . . . 11.0 x 1.0-in vented disc
R: . . . . 11.1 x 0.5-in disc
Power assist . . . . vacuum with anti-lock control

WHEELS AND TIRES
Wheel size . . . . 7.0 x 15 in
Wheel type . . . . cast aluminum
Tires . . . . Pirelli P600, 205/55VR-15
Test inflation pressures, F/R . . . . 31/34 psi

CAR AND DRIVER TEST RESULTS

ACCELERATION
	Seconds
Zero to 30 mph	2.4
40 mph	3.7
50 mph	5.1
60 mph	6.9
70 mph	9.1
80 mph	11.8
90 mph	14.6
100 mph	19.6
110 mph	25.3
Top-gear passing time, 30–50 mph	9.4
50–70 mph	9.5
Standing ¼-mile	15.2 sec @ 92 mph
Top speed	141 mph

BRAKING
70–0 mph @ impending lockup . . . . 179 ft
Fade . . . . **none** moderate heavy

HANDLING
Roadholding, 300-ft-dia skidpad . . . . 0.81 g
Understeer . . . . minimal **moderate** excessive

COAST-DOWN MEASUREMENTS
Road horsepower @ 30 mph . . . . 5 hp
50 mph . . . . 13 hp
70 mph . . . . 29 hp

FUEL ECONOMY
EPA city driving . . . . **17 mpg**
EPA highway driving . . . . 29 mpg
C/D observed fuel economy . . . . **20 mpg**

INTERIOR SOUND LEVEL
Idle . . . . 59 dBA
Full-throttle acceleration . . . . 82 dBA
70-mph cruising . . . . 72 dBA
70-mph coasting . . . . 71 dBA

CURRENT BASE PRICE dollars x 1000

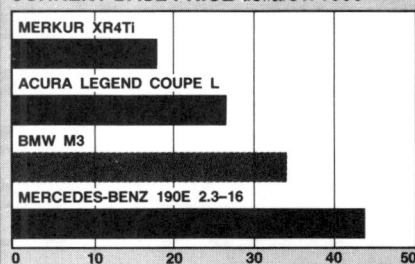

MERKUR XR4Ti
ACURA LEGEND COUPE L
BMW M3
MERCEDES-BENZ 190E 2.3–16

0 10 20 30 40 50

ACCELERATION seconds

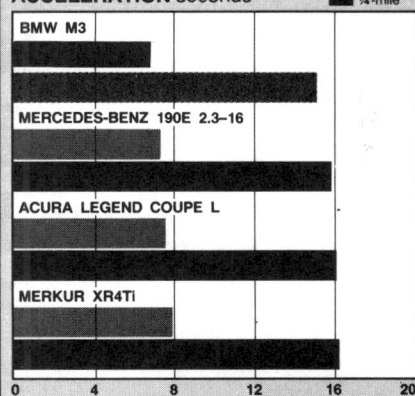

■ 0–60 mph
■ ¼-mile

BMW M3
MERCEDES-BENZ 190E 2.3–16
ACURA LEGEND COUPE L
MERKUR XR4Ti

0 4 8 12 16 20

70–0 MPH BRAKING feet

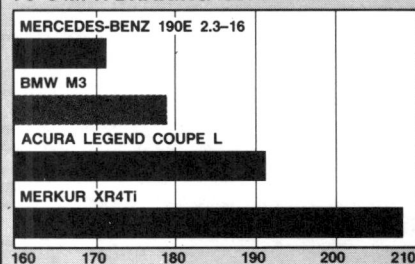

MERCEDES-BENZ 190E 2.3–16
BMW M3
ACURA LEGEND COUPE L
MERKUR XR4Ti

160 170 180 190 200 210

ROADHOLDING 300-foot skidpad, g

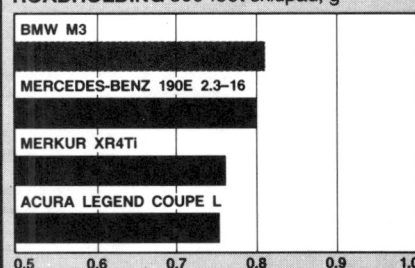

BMW M3
MERCEDES-BENZ 190E 2.3–16
MERKUR XR4Ti
ACURA LEGEND COUPE L

0.5 0.6 0.7 0.8 0.9 1.0

EPA ESTIMATED FUEL ECONOMY mpg

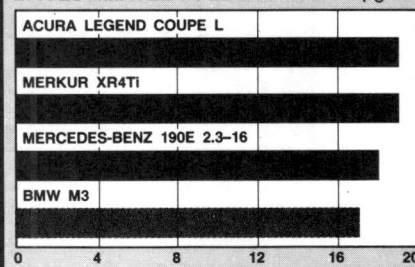

ACURA LEGEND COUPE L
MERKUR XR4Ti
MERCEDES-BENZ 190E 2.3–16
BMW M3

0 4 8 12 16 20

Offering supercar performance with four-door comfort, this hyper-sports model is in a league of it's own. With a top speed of 250 km/h, it's the fastest sedan we've ever tested...

BMW M5

SOUTH African performance car fans are among the most pampered in the world. Relative to our country's size, we have a fantastic array of potent machinery to pepper our shopping lists, and for BMW's many adherents here, the choice is mouthwatering.

There's the exacting 325i and 325i "S" for the seat-of-the-pants types who put agility before rear passenger space. For the company top-dog with fire in his

KEY FIGURES

Maximum speed	250 km/h
1 km sprint	26,94 seconds
Fuel tank capacity	70 litres
Litres/100 km at 100	9,18
Optimum fuel range at 100	762 km
*Fuel index	12,85
Engine revs per km	1 563
National list price	R109 970

(*Consumption at 100 plus 40%)

heart there is the 745i, a limousine that has almost supercar get-up-and-go. And now, for the man who wants the fastest locally-produced full-sized sedan (and who can afford to pay for it) there is the M5. A simple title for a simple concept: the meanest, quickest, most enthralling car ever unleashed by BMW SA.

To give M5 owners an idea of how lucky they are, it should be noted that Britain has only been allocated 200 of these cars a year, while BMW SA is getting 100 a year, even though South Africa's new car market is only a fraction of the UK's in size. Moreover, the M5 sold here is the most up-market spec model available anywhere and European and British owners have to specially order all the BMW Motorsport division parts that are automatically ear-marked for "our" M5s, if they want them. The catch is that a South African M5 costs well over R100 000, but once you've run

your eye down the vital statistics the shock value of that price tag tends to diminish.

When the average motorist views the M5 for the first time, there are two things he wants to know: "How fast does she go, mister, and — *by the way* — how much does it cost?" It is difficult to judge which answer causes the eyebrows to elevate more — the top speed or the price — but it's probably the latter, mainly because very few people can really relate to a speed of 250 km/h.

That's right; on a level road, with two adults on board, an almost-full tank of fuel and just a whisper of a breeze in the air, our test-model BMW M5, with a mass of 1 580 kg excluding the drivers, cut through the atmosphere at a genuine 250 km/h. And at top speed, with the accurate rev-counter reading a shade over 6 500 and the speedo needle nudging 270, the big Beemer tracks straight and true, with just a touch of bucking on the sports-style leather-rimmed steering wheel to let you know that you are part of an exceptional occurrence. To put the car's speed capabilities into perspective, we checked on the maxima achieved by other roadburners and after that, we were even more impressed.

That 250 top-end makes the M5 the fastest sedan and the fastest locally built car we've ever tested. It's the third fastest overall, putting it in company with the Porsche 944 Turbo (256 km/h), the Porsche 928S A/T (253), the Porsche 911 Carrera (248) and an (imported) Mercedes-Benz 560 SEC (247).

Visually, the South African M5 is the most exciting of its type available anywhere. Whereas the European models are fitted with 535i-type wheels and a minimum of trim changes, the models produced here uses the massive 7,5J 16 inch alloy rims, front and rear spoilers and the side skirts from BMW's Motorsport division, to create a radically different-looking pursuit missile.

The engine fitted to the M5 is quite different from that used in the 535i.

For the ultimate in powerlines we recommend the M5 — supercar performance in a four-door package, with superb body detailing and road wheels setting it apart from other 5-Series BMWs.

Starting with the fully electronic Bosch Motronic fuel injection system, the induction tracts are of a shorter design for more top-end power, with each tube drawing air directly from a large airbox. The cylinder head features a beautifully-finned alloy cam cover, with an "M Power" logo indicating the serious nature of this device, which has its origins in BMW's supercar of the early 1980s, the M1.

The twin overhead camshafts activate the 24 valves directly, through shimmed bucket-tappets instead of rockers, as used in the s-o-h-c 535i motor. Interestingly, the cam lobe profiles allow for slightly less lift on the M-power 24-valve motor than those used on the 12-valve 535i unit, but obviously with 12 extra valves in play, the flow characteristics are vastly improved.

Special forged pistons raise the compression ratio to 10,5:1 (10:1 on the 535i) and specially-made stronger conrods keep the act together at maximum revs, in conjunction with a steel crank re-worked by the Motorsport crew. It is comforting to note that the motor's standard bottom end can take 8 500 revs with no sign of pain, according to local track ace Tony Viana, but in the interests of longevity, a fuel cut-out has been programmed into the engine management system at 6 900 r/min.

The cylinder geometry is also different, for while the race-bred powerplant has a 93,4 mm bore and an 84 mm stroke, the 535i's is less oversquare at 92/86.

210 KW AT 6 500 R/MIN

Add a fabricated set of curling exhaust headers and a big-bore, minimally restrictive pipe and silencer system and you are looking at a power output of 210 kW at a suitably elevated 6 500 r/min, with peak torque of 340 N.m being delivered at 4 500.

Even with a quoted curb mass of 1 502 kg and fairly indifferent aerodynamics, this kind of powerplay delivers the goods in no uncertain fashion, as evi-

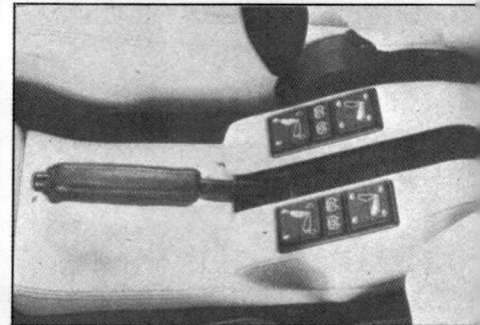

denced by our impressive sprint times. Driven with no regard for tyre wear or synchromesh, the M5 storms through to 100 km/h from rest in a mere 6,75 seconds, needing only one gearchange to reach this mark. Again, this is Porsche territory that the BMW is exploring, and the standing kilometre sprint takes 26,94 seconds while the 944 Turbo's time for the same distance is 26,71 seconds.

The strange thing is that initially, the M5 just doesn't feel *that* fast. Despite the massive power available practically anywhere in the power band, the straight-six 24-valver is still relatively peaky and there are notable "steps" in the power delivery at 5 000 and again at 5 900, after which comes another big surge towards the summit.

The incremental nature of the power delivery only becomes noticeable when you're engaged in white-knuckle, foot-on-the-floor motoring; otherwise it is all silky smoothness, with the engine pulling shudder-free from 1 400 r/min in fifth gear. Another surprising aspect to the car, for those versed on BMW lore, is that the five-speed manual gearbox has a conventional pattern, instead of the "dog-leg first gear" arrangement found on some other sporting BMWs.

The 210 kW d-o-h-c six-cylinder engine is a work of art — every cable and casting designed to look good as well as to be extremely efficient (opposite page). The same can be said for the interior (above) finished in white leather, with the front seats electrically adjustable to almost any configuration, using the "illustrated buttons" on either side of the handbrake lever. Front seats grip particularly firmly (top, left) and the instrumentation is refined (right) and virtually flawless.

The M5 uses a specially robust Gëtrag 'box with overdrive fifth gear (0,81:1), coupled to a 3,73:1 final drive in the rear axle, which has a limited slip differential, with a 25 per cent lock.

According to BMW SA's development department, the decision to use the conventional-shift gearbox was made in Germany when it was discovered that the "sports" 'box would be overstressed when mated to 210 kW of power, or at least would not allow a sufficient margin for reliability. Our test team welcomed this move as we have never really liked the "sports" 'box, feeling that it required too much concentration for swift, sporting driving.

CONNOISSEUR'S DELIGHT

Wringing the best out of the M5 over a challenging road has to be one of the finest automotive experiences available in this country. The cockpit is a connoisseur's delight, and the test car interior was upholstered in striking white leather with impeccable attention to detail.

The front seats are electrically adjustable to suit practically every physique and taste, including a tilt adjustment for the front squab, which we extended to its limits to enable a tallish driver to see the upper segment of the speedometer beneath the thick, leather-rimmed sports steering wheel. The steering column also has a telescopic adjustment which we made use of to overcome the speedo obscurement that has been a long-standing fault of 5-Series BMWs.

Apart from the luxurious aura created by the white leather — which is sensitive to scuffing and may not look so appealing after a few years use — creature comforts are well-catered for. The (optional) Becker Grand Prix stereo system is one of the best in the business and there are air-conditioning, electric windows and electrically-adjustable wing mirrors. An electric sunroof is an option.

For those who don't enjoy air-conditioning, the five-vent airflow system is notably efficient, aided by a quiet fan and with quick-acting demister nozzles for the windscreen and side-windows.

In view of all these efforts to ensure that M5 owners need never soil their handkerchiefs on wet winter mornings, we were surprised that the headlights were not fitted with spray nozzles and wipers. Perhaps it wasn't possible with the round configuration of the four halogen lights, although with the addition of fog-lamps integrated into the front air-dam, night time vision is not a problem.

In fact we rate the lights as one of the car's most impressive safety features, along with the heavily re-inforced body, the impressive handling and the brakes. The M5 comes standard with ABS control of the diagonal dual-circuit, all-disc brake system, which features fixed four-pot calipers for the front ventilated discs and fist-type calipers for the solid ones at the rear.

SUPERB BRAKING

The brakes are among the best we've sampled in any car and the way they slow this heavy saloon down from very high speeds is truly astonishing. In our 10-stop braking test from 100 km/h we recorded an average stopping time of 3,0 seconds with a best stop breaking the magical three-second barrier at 2,9 seconds and the worst stop at 3,1 seconds, showing the remarkable consistency of the set-up.

No doubt the fat, sticky Pirelli P700s, of 225/50 VR 16 configuration, play their part, as they do in the car's handling and road-holding. The M5 feels agile and nimble as no other 5-Series BMW, thanks to some subtle refinements to suspension damping, geometry and the wheel/tyre combination.

Although suspension geometry remains basically the same, a few more minutes of negative camber have been dialled into the front end, while at the rear, the wheels' camber is set 12 minutes more positive than those on the 535i, which

keeps them slightly more vertical during hard-cornering. The springs fitted front and rear are stiffer and shorter, reducing the ride height slightly, and the Bilstein dampers have stiffer rebound settings.

Responding to its small-diameter steering wheel and variable-with-engine-speed power assistance, the M5 turns into a corner quickly but with no trace of nervousness, the communication between road wheels and the driver's palms being direct and finely detailed. At no time in the M5 did we encounter the tail-happiness found in the previous generation 535 and in fact the M5 is set up to understeer just a touch before a well-signalled transition to oversteer, at the giddy limits of adhesion.

We use the term ''giddy'' because you would have to be a little crazy to break the tyres loose in the first place, but after a trip or two to The Edge we are happy to report that the M5 is easily brought back into line. In fact, we noted that the once-dreaded trailing throttle oversteer is also a thing of the past on this model, so you can take your pick in controlling a rear-end step-out by either staying on the power and dialling in some opposite lock, or by simply backing-off the throttle a touch. M5 owners should be warned, however, that things happen very quickly at this stage of the proceedings, simply because you have to travel so fast to provoke the slide in the first place.

Driven in sedate fashion the M5 can be driven smoothly, but there is always a sense of urgency about the plot, partly because of the enormous power always on

tap but also because the clutch is very heavy, and making smooth changes takes a bit of practice. Driven in this fashion, the M5 returns a consumption figure of around 12 to 13 litres/100 km as our fuel index figure indicates, but with hard driving the consumption increases to about 15 or 16 litres/100 km — still very acceptable for a car of this nature.

When driven at moderate speeds the ride is acceptably comfortable, with a slight trace of bump-thump from the ultra-low profile Pirellis but nothing that would have you biting your tongue accidentally. In the interests of theft protection the windows are etched with serial numbers and the door lock has an additional key position that prevents the interior door knobs from being pulled into the un-locked position, should a thief gain access through a window. Imagine trying to explain to a policeman why you prefer to enter and exit your car through the window, just like the NASCAR stock car drivers do!

TEST SUMMARY

It is difficult to criticise the M5 as the car is so beautifully built, is so fast, behaves impeccably in fast and slow corners and has a lot of image. But despite all its attributes, it still falls short of creating the kind of excitement that a real supercar like a Porsche 911 or a Ferrari 308 can give, possibly because it *is* so refined. Yet for what it offers, supercar-like performance with four-door comfort for four or even five adults, it is in a league of its own. ●

SPECIFICATIONS

ENGINE:
Cylinders	six in-line
Fuel supply	Bosch Motronic fuel injection
Bore/stroke	93,4/84,0 mm
Cubic capacity	3 453 cm^3
Compression ratio	10,5 to 1
Valve gear	d-o-h-c, four valves per cylinder
Ignition	electronic
Main bearings	seven
Fuel requirement	98-octane Coast, 93-octane Reef
Cooling	water

ENGINE OUTPUT:
Max. power I.S.O. (kW)	210
Power peak (r/min)	6 500
Max. usable r/min	6 900
Max. torque (N.m)	340
Torque peak (r/min)	4 500

TRANSMISSION:
Forward speeds	five
Gearshift	console
Low gear	3,51 to 1
2nd gear	2,08 to 1
3rd gear	1,35 to 1
4th gear	1,00 to 1
Top gear	0,81 to 1
Reverse gear	3,71 to 1
Final drive	3,73 to 1
Drive wheels	rear

WHEELS AND TYRES:
Road wheels	alloy
Rim width	7,5J
Tyres	225/50 VR 16
Tyre pressures (front)	250 to 270 kPa
Tyre pressures (rear)	270 to 330 kPa

BRAKES:
Front	ventilated discs, four piston calipers
Rear	discs and separate handbrake drums
Hydraulics	ABS, dual circuit diagonally split
Boosting	hydraulic pump
Handbrake position	between seats

STEERING:
Type	ZF ball and nut, power assisted
Lock to lock	4,6 turns
Turning circle	10,4 metres

MEASUREMENTS:
Length overall	4 620 mm
Width overall	1 700 mm
Height overall	1 400 mm
Wheelbase	2 625 mm
Front track	1 430 mm
Rear track	1 465 mm
Ground clearance	100 mm
Licensing mass	1 502 kg
Mass as tested	1 580 kg

SUSPENSION:
Front	independent
Type	MacPherson struts, lower wishbones, trailing links, stabiliser bar, gas shock absorbers
Rear	independent
Type	dual jointed half axles, spring struts, semi trailing arms

CAPACITIES:
Seating	4/5
Fuel tank	70 litres
Luggage trunk	320 dm^3

WARRANTY:
12 months irrespective of distance.

TEST CAR FROM:
BMW South Africa.

ACCELERATION

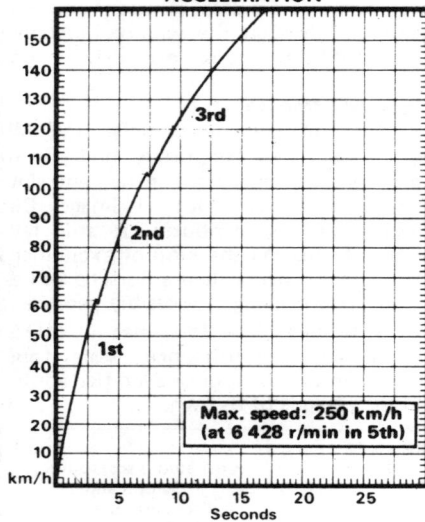

Max. speed: 250 km/h
(at 6 428 r/min in 5th)

PERFORMANCE FACTORS:
Power/mass (W/kg) net . . 139,80
Frontal area (m²) 2,38
km/h per 1 000 r/min (top). 38,39
(Calculated on licensing mass, gross frontal area, gearing and I.S.O. power output)

TEST CONDITIONS:
Altitude . . . . at sea level
Weather . . . . .warm, light wind
Fuel used. 98-octane
Test car's odometer 5 627

GRADIENT ABILITY

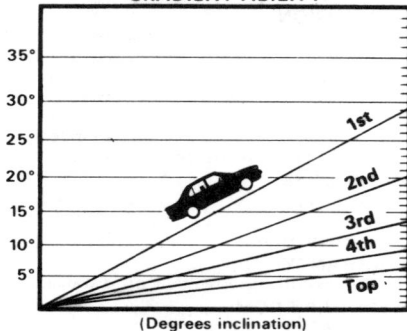

(Degrees inclination)

MAXIMUM SPEED (km/h):

True speed 250
Speedometer reading 270
Calibration:

Indicated:	60	80	100	120
True speed:	56	75	96	113

ACCELERATION (seconds):

0-60. 3,14
0-80. 4,95
0-100. 6,75
0-120 9,44
1 km sprint 26,94
Terminal speed . . . . . .199,7 km/h

OVERTAKING ACCELERATION:

	3rd	4th	Top
40-60	3,27	4,61	6,56
60-80	2,71	4,54	6,19
80-100	2,62	4,06	6,68
100-120	2,51	4,03	6,28

FUEL CONSUMPTION (litres/100 km):

60. 7,24
70. 7,67
80. 8,12
90. 8,61
100 9,18
110 9,80
120 10,60

BRAKING TEST:

From 100 km/h
Best stop 2,9
Worst stop 3,1
Average 3,0
(Measured in seconds with stops from true speeds at 30-second intervals on a good bitumenised surface)

GRADIENTS IN GEARS:

Low gear . . . . 1 in 1,8
2nd gear . . . . 1 in 2,7
3rd gear . . . . 1 in 4,2
4th gear . . . . 1 in 6,2
Top gear . . . . 1 in 9,1
(Tabulated from Tapley (x gravity) readings, car carrying test crew of two and standard test equipment)

GEARED SPEEDS (km/h):

Low gear . . . . 58* . . . 62
2nd gear . . . . 97* . . . 105
3rd gear . . . . .150* . . 161
4th gear . . . . .202* . . 218
Top gear . . . . .250* . . 269
(Calculated at engine power peak* — 6 500 r/min and at max. usable r/min — 7 000)

INTERIOR NOISE LEVELS:

	Mech.	Wind	Road
Idling	.48	—	—
60	.61	—	—
80	.64	70	68
100	.67	74	70

(Measured in decibels, "A" weighting, averaging runs both ways on a level road: "mechanical" with car closed; "wind" with one window fully open; "road" on a coarse road surface)

ENGINE SPEED

Max. torque: 4 500 r/min

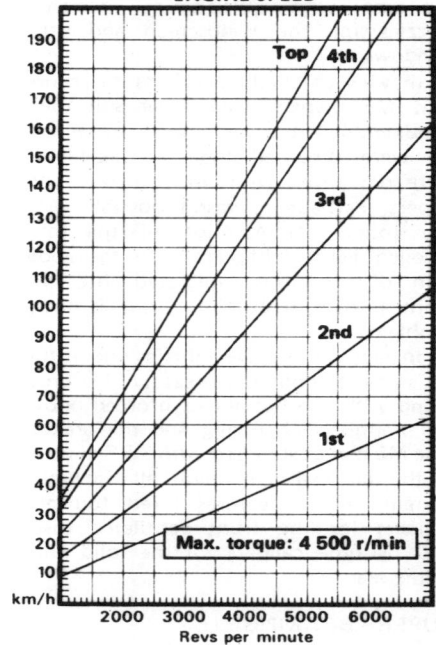

BRAKING DISTANCES

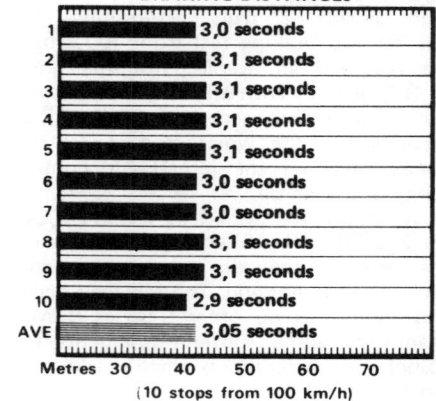

1		3,0 seconds
2		3,1 seconds
3		3,1 seconds
4		3,1 seconds
5		3,1 seconds
6		3,0 seconds
7		3,0 seconds
8		3,1 seconds
9		3,1 seconds
10		2,9 seconds
AVE		3,05 seconds

Metres 30 40 50 60 70
(10 stops from 100 km/h)

BMW M5

Getting that old-time hot-rod religion.

• Charles Darwin died in 1882, four years too early to contribute anything to automobile engineering, but he was hell on wheels when it came to his theory of evolution. Almost single-handedly he threw a monkey wrench into by–the–Good Book, old-time religion. Roughly put, Darwin held that God didn't really pop packets of Instant Adam 'n' Eve into his celestial microwave; instead, the subhuman race took time to crawl before it could walk, and somewhere along the line we *Homo sapiens* all had apes swinging in our family trees. Now here we are, naturally selected simian descendants—smack in the age of manna, Vanna, and hot-car nirvana—and those of us who like to monkey around with cars have got the good life. For proof that automotive evolution can be nearly as miraculous as God's more basic monkey business, you have only to go out and strap on the new BMW M5.

Webster's definition of Darwinian theory concludes "that the forms which survive are those that are best adapted to the environment." As BMW knows, survival in today's auto environment calls for power, and lots of it—thanks largely to cheap gas, and lots of it. Under the M5's hood rages a twin-cam, 24-valve, 3.5-liter, six-cylinder ghost of an engine. Resurrected from the fabulous mid-engined M1 coupe of the late seventies, it whirs with 256 horsepower in its new home. Hand-assembled by BMW's Motorsport branch, the big six hurls the meaty-tired, big-braked, tautly suspended M5 brickbat to almost 150 mph. As a prime example of the high-performance roadware thundering through our times, the M5 proves that, Darwin and your loan officer notwithstanding, now is the age to go ape.

So what if the M5 looks as if it were designed when Darwin was still living? Bimmer buyers have naturally selected this shape as one that strikes their fancy. BMW roughed in the profile of its second-generation 5-series sedan back in the late seventies. In those days, the factory was determined not to digress from its familiar blocky styling, and the four-door's contours came out like almost a genetic dupli-

cate of its predecessor's.

In 1982, BMW delivered its new box to America as the 528e. Its low-revving, 2.7-liter engine paid homage to fuel economy and low-end torque, undercutting BMW's reputation as a builder of "ultimate driving machines." While Europe continued to enjoy the output of BMW's horsepower department, American Bimmer loyalists were forced into the slow lane. BMW's U.S. sales continued to set records, however, as the company coasted on an image built on fifteen years of rave reviews.

Then, in 1983, a new evolutionary form emerged: Audi's slick 5000S blasted out of the wind tunnel and threatened to show the rest of the world's sedans just who was the fittest of them all. Not only was the trimly rounded Audi the first modern sedan to manifest serious attention to aerodynamics, but its creators quickly backed up their threat with turbocharging and four-wheel drive.

BMW, faced with this triple whammy of technology, did what it always does, sooner or later, in the face of adversity: it dug deeper into its power bin. Much deeper. Along came the 533i and the 535i, the first "i" cars to suggest that BMW was here to play for keeps. This year came the hard cases from the Motorsport mob, finally bringing us the same good stuff that Europe has been taking for granted. The M6

coupe (*C/D*, July) lit the way with a top speed of 144 mph. The mini-motor M3 pocket rocket (*C/D*, November) ripped right up to 141. And now comes the M5, denying its four-door demeanor by booming into battle at 147 mph.

The M5 and the M6 share the same engine: 3.5 liters of displacement, Bosch Motronic fuel injection, an aluminum crossflow head, four valves per cylinder, machined intake and exhaust ports, pent-roof combustion chambers, a 9.8:1 compression ratio, an oil cooler, a low-restriction catalyst, dual exhausts—and 256 hp from 211 cubic inches. BMW's biggest six displaces marginally more than its single-overhead-cam sister, and thanks in part to a larger bore and a shorter stroke, it revs higher. In any of the first four gears of the Getrag five-speed, the M motor flies past its 6500-rpm power peak to a 6900-rpm redline. Then a quick double snick of the gearbox pumps the big six back into the heart of its broad power band, and the lusty *vroom* continues. (BMW offers no automatic to drag down the M5's output at the rear wheels.) Our fifth wheel translates the 3504-pounder's acceleration into a 0-to-60 time of 6.3 seconds. The M5 covers the quarter-mile in 14.6 seconds, crossing the line at 95 mph, with another 52 mph still to come. For a boxed-off, four-door folks-wagon, those are hot numbers. The only things that can cool them are a smallish gas tank and a 10-mpg EPA city fuel-economy rating. Happily, as hard as we hammered the M5, we averaged a more reasonable 15 mpg.

Black, and only black, smothers the M5. From paint to trim, BMW allows no less serious body color to bear false witness to its intent. Other than its deadly coloring, its thickset stance, and its add-on aerodynamic aids, the M5's only tip-offs are flashy blue-purple-and-red-banded Motorsport badges on the grille and the tail. The spider-web cast-aluminum wheels shine in rich silver. The bodywork wears an aggressively ducted air dam up front and a rubber-ribbed, stylized-wickerbill spoiler at the rear. The only awkward note is the U.S.-spec bumpers poking out defiantly at both ends, only partly compensating for their clumpy appearance with excellent 5-mph impact ratings.

The Motorsport mavens had pavement abuse in mind when they engineered the M5's chassis. Their first act was to specify a great, gummy Pirelli P700 for each wheel well. Mounted on a 7.5-by-16-inch

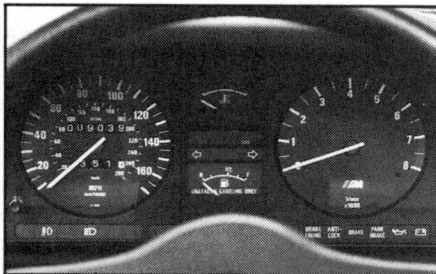

wheel, each 225/50VR-16 tire squeezes into its standard 5-series fender arch like a linebacker's neck through a pipsqueak's collar. Outfitted thus, the M5 abuses the skidpad up to an outstanding and easily controllable cornering limit of 0.83 g.

BMW thoughtfully provides two major handling aids—one complex, one simple. First, as in other contemporary Bimmers, the patented Track Link suspension arrangement cancels any latent lift-throttle-oversteer tendencies from the semi-trailing-arm rear suspension. Second, conservative tire-pressure recommendations—36 psi in front, 40 in the rear—add another dose of understeer. (We found better results on the road with equal front and rear inflation.)

Thanks to the taut reflexes of the M5's steering and suspension, any pavement is open to abuse. The anti-roll bars remain unchanged from 535i specs, but shorter progressive-rate springs and heftier gas-pressure front struts and rear shocks encourage hard driving without fear of nasty repercussions. The damping calibration swiftly soaks up problems in one efficient cycle of motion. It's a firm cycle, but whether you're sightseeing or running hard, the M5 gives good control, never threatening to make a monkey of you.

The M5 is so quick that waiting to pass someone safely creates no frustration: you feel you can afford good traffic manners because the machine quickly compensates for any delays. Its behavior is so calmly composed, so safe and stable at the elevated speeds it readily attains, that in a strange way it calls for added caution: you have to be constantly mindful that trouble

can leap up around you too fast for human reaction times to handle.

The M5's sizable, ABS-outfitted, four-wheel disc brakes, which are vented in front, always do their best to keep you from harm. Bosch's electronic anti-lock circuitry never interferes with BMW's firm braking action, even during hard driving. Yet it stands ever ready should you need to stand on the pedal in an emergency. With the help of the fat Pirellis, the ABS stops the M5 from 70 mph in only 166 feet. This ranks second by only two feet to the modern *C/D* record, set by the Corvette and the Porsche 928S4.

Inside, the M5 is laid out better than the Corvette, but perfection is a detail or two away. Unless you get lucky with BMW's complicated power-seat buttons, you may have to search repeatedly for the right

seating position. And there are no memory buttons to help. Luckily, the steeply raked steering column allows you to telescope the padded sport wheel. When your reach finally takes the proper measure of the controls, you find pedals perfectly beneath your feet. Effortless heel-and-toe

action and the precise give-and-take of every control ensure that neither car nor driver feels out of phase.

The M5's insides look like the finely instrumented and tailored pilots' cocoon of a deep-space launch vehicle. Judging by the expanses of creamy leather, several cows were generous enough to give the hides right off their backs. The sport seats look great and grab like baseball mitts, but the seatbacks' bulging Motorsport badges tend to gouge the backs of tall travelers. In addition, the tricolor tape that decorates the badges has a tendency to curl at the edges. The dash boasts a broad array of levers, buttons, and dials whose effects on the M5's climate-control and premium sound systems are top-notch. Alas, our test car's intermittent-wiper setting failed to intermit. And the wiper-control lever

was too easily bumped into action when we keyed the ignition.

If these few imperfections put you off the M5, you're reading the wrong periodical. This battling BMW is one of those rare, fine cars that have grit—loads of it. And next year, despite the recent infusion of speed and character, BMW will introduce an even more highly evolved 5-series. It will be shaped more like an egg than a crate, and the M5 that will eventually descend from the new platform will no doubt have a friendlier relationship with the wind.

Sixty years ago, the shape of Darwin's enlightening new ideas frightened Bible thumpers into trying to oust his theory from the schools. In 1960, a great film called *Inherit the Wind* dramatized the infamous "Tennessee monkey trial" that resulted. The actor Spencer Tracy, restating Clarence Darrow's case against ignorance, turned the Bible on its thumpers and quoted, "He that troubleth his own house shall inherit the wind." Tracy/Darrow lost the case, but in the end Darwin's theory won the day, just as it has again with BMW's evolutionary return to its high-performance origins.

The next M5 will be even better, even faster. The wind won't have a hope in hell of keeping up. But you can already find ample enlightenment in BMW's book of Motorsport. Buy this M5 and you won't have to wait to inherit the wind: you can go out and clobber it into submission right now. —*Larry Griffin*

COUNTERPOINT

• I love the split personality of the BMW M5. From the outside, this car is all business: no gimmicks, no screaming power emblems, no wings. Just the stoic stance of an average, everyday German sedan.

Climb in and turn the key, though, and the M5 is instantly transformed from Dr. Jekyll into Mr. Run-Away-and-Hyde. Engine, engine, engine—the key word here is "engine." Punch the M5's lusty, 24-valve six, and you'll leave behind forever the mundane world of the everyday sedan. Should the engine's scintillating performance not wake you from your staid-sedan slumber, its unearthly shriek certainly will. Unless you regularly strap yourself into an F1 car, you're unlikely to know such mechanized musical splendor.

Of course, the M5 is outrageously expensive. But then, it wasn't designed to lead the price brigade. The M5 is a no-compromises, foot-to-the-floor screamer built for those who demand the ultimate in speed and refinement. The few who can afford it are going to have a ball. —*Arthur St. Antoine*

Can you believe what BMW is up to? The tightly laced Munich firm is riling Mercedes with its twelve-cylinder 750iL, raising havoc with aftermarket tuners by launching one M-machine after another, and kicking dirt on Porsche by announcing that the Z1 roadster is a go program. What's next, a $4000 BMW to take on the Koreans?

Probably not. With the M-class, BMW is well and truly back on the wavelength that established its reputation in the first place. The M5 is the classic executive express: patently practical, faster than a speeding 560SEL, more macho than an East L.A. lowrider with a

hyperactive suspension.

I'd rush right out and buy an M5 if not for one niggling problem: you don't get much change back from $50,000 when you plunk down funds for the fastest Bimmer in all the land. But BMWs have always been expensive, and those of you with looser purse strings shouldn't fret over spending a little extra for Bavaria's performance flagship. At a buck a thrill, you'll write the investment off in no time. —*Don Sherman*

Big news here: the M5 is not a car. Okay, I know it has tires, an engine, and a steering wheel. But if you start thinking like that, you're gonna toss this magazine down and accuse me of getting silly about yet another overpriced Teutonic road bomber.

Yes, you can find the same basic pieces—a 24-valve six, an independent suspension, and a five-speed—in less expensive sedans, like the Acura Legend and the Sterling 825. Fine transportation devices, those, but the M5 is supposed to do more than just move you. Its mission is to *move* you.

That it does. You buy this car for its soul. Everything about it oozes confidence. It's got the heart of a tiger. The big six sounds as if it could rip a V-8 to shreds, and it feels that way, too. The bespoilered bodywork gives off all the right messages.

Listen, a guy I know just bought himself an M-car—an M6 actually, but no matter. He's bright, witty, a knowledgeable enthusiast, a fine race driver, and successful enough that he doesn't have to worry about the price of the reward he's given himself. He loves his new toy. Judging the M5 as a price-is-no-object toy, I love it, too. —*Rich Ceppos*

Vehicle type: front-engine, rear-wheel-drive, 5-passenger, 4-door sedan

Price as tested: $48,470

Options on test car: base BMW M5, $45,500; heated front seats, $200; gas-guzzler tax, $2250; freight, $520

Standard accessories: power steering, windows, seats, locks, and sunroof, A/C, cruise control, rear defroster

Sound system: BMW AM/FM-stereo radio/cassette, 8 speakers

ENGINE
Type .... 6-in-line, iron block and aluminum head
Bore x stroke .... 3.68 x 3.31 in, 93.4 x 84.0mm
Displacement .... 211 cu in, 3453cc
Compression ratio 9.8:1
Engine-control system .... Bosch Motronic with port fuel injection
Emissions controls .... 3-way catalytic converter, feedback fuel-air-ratio control
Valve gear .... chain-driven double overhead cams, 4 valves per cylinder
Power (SAE net) .... 256 bhp @ 6500 rpm
Torque (SAE net) .... 243 lb-ft @ 4500 rpm
Redline .... 6900 rpm

DRIVETRAIN
Transmission 5-speed
Final-drive ratio .... 3.91:1, limited slip

Gear	Ratio	Mph/1000 rpm	Max. test speed
I	3.51	5.3	36 mph (6900 rpm)
II	2.08	8.9	61 mph (6900 rpm)
III	1.35	13.7	95 mph (6900 rpm)
IV	1.00	18.5	128 mph (6900 rpm)
V	0.81	22.8	147 mph (6450 rpm)

DIMENSIONS AND CAPACITIES
Wheelbase .... 103.3 in
Track, F/R .... 56.3/57.7 in
Length .... 189.0 in
Width .... 66.9 in
Height .... 55.7 in
Ground clearance .... 4.9 in
Curb weight .... 3504 lb
Weight distribution, F/R 52.2/47.8%
Fuel capacity .... 16.6 gal
Oil capacity .... 6.1 qt
Water capacity .... 12.7 qt

CHASSIS/BODY
Type .... unit construction with one rubber-isolated crossmember
Body material .... welded steel stampings

INTERIOR
SAE volume, front seat .... 49 cu ft
rear seat .... 37 cu ft
trunk space .... 14 cu ft
Front seats .... bucket
Seat adjustments .... fore and aft, seatback angle, front height, rear height, thigh support
General comfort .... poor fair **good** excellent
Fore-and-aft support .... poor fair good **excellent**
Lateral support .... poor fair good **excellent**

SUSPENSION
F: .... ind, strut located by one trailing link and one lateral link, coil springs, anti-roll bar
R: ...... ind, semi-trailing arm, coil springs, anti-roll bar

STEERING
Type .... recirculating ball, power-assisted
Turns lock-to-lock 3.5
Turning circle curb-to-curb .... 32.8 ft

BRAKES
F: .... 11.8 x 1.2-in vented disc
R: .... 11.2 x 0.4-in disc
Power assist .... hydraulic with anti-lock control

WHEELS AND TIRES
Wheel size .... 7.5 x 16 in
Wheel type .... cast aluminum
Tires .... Pirelli P700, 225/50VR-16
Test inflation pressures, F/R .... 36/40 psi

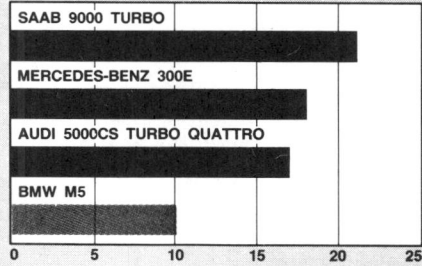

CAR AND DRIVER TEST RESULTS

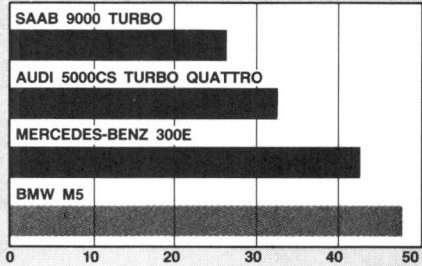

ACCELERATION
	Seconds
Zero to 30 mph	2.1
40 mph	3.3
50 mph	4.7
60 mph	6.3
70 mph	8.3
80 mph	10.7
90 mph	13.0
100 mph	17.3
110 mph	23.2
Top-gear passing time, 30–50 mph	9.3
50–70 mph	9.8
Standing ¼-mile	14.6 sec @ 95 mph
Top speed	147 mph

BRAKING
70–0 mph @ impending lockup .... 166 ft
Fade .... **none** moderate heavy

HANDLING
Roadholding, 300-ft-dia skidpad .... 0.83 g
Understeer............. minimal **moderate** excessive

COAST-DOWN MEASUREMENTS
Road horsepower @ 30 mph .... 6 hp
50 mph .... 15 hp
70 mph .... 33 hp

FUEL ECONOMY
EPA city driving .... **10 mpg**
EPA highway driving .... 19 mpg
C/D observed fuel economy .... **15 mpg**

INTERIOR SOUND LEVEL
Idle .... 54 dBA
Full-throttle acceleration .... 77 dBA
70-mph cruising .... 71 dBA
70-mph coasting .... 69 dBA

CURRENT BASE PRICE dollars x 1000
- SAAB 9000 TURBO
- AUDI 5000CS TURBO QUATTRO
- MERCEDES-BENZ 300E
- BMW M5

ACCELERATION seconds
0–60 mph / ¼-mile
- BMW M5
- SAAB 9000 TURBO
- MERCEDES-BENZ 300E
- AUDI 5000CS TURBO QUATTRO

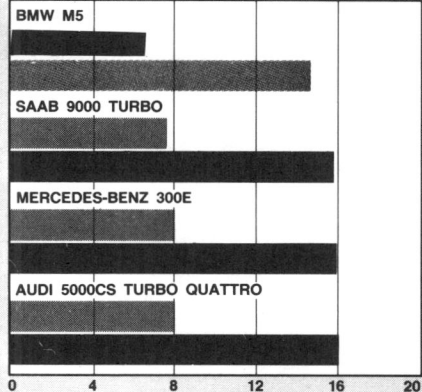

70–0 MPH BRAKING feet
- BMW M5
- AUDI 5000CS TURBO QUATTRO
- MERCEDES-BENZ 300E
- SAAB 9000 TURBO

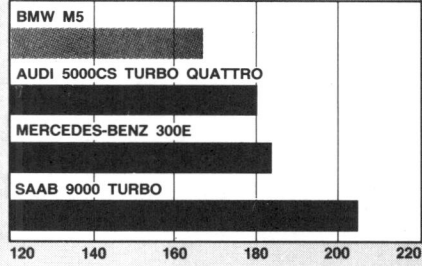

ROADHOLDING 300-foot skidpad, g
- BMW M5
- MERCEDES-BENZ 300E
- AUDI 5000CS TURBO QUATTRO
- SAAB 9000 TURBO

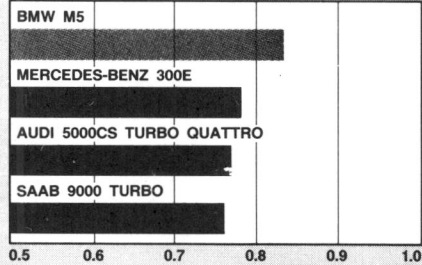

EPA ESTIMATED FUEL ECONOMY mpg
- SAAB 9000 TURBO
- MERCEDES-BENZ 300E
- AUDI 5000CS TURBO QUATTRO
- BMW M5

REVIEW

PASSAGES

BMW's hot M3 takes off as M-B's 2.3-16 fades away.

BY KEVIN SMITH

Ann Arbor—What we have here is a story about a beginning and an ending. Think in terms of yin and yang, birth and death, the very rhythms of the cosmos. Think about the passing of a grand baton.

Or just think about exactly what's happening. An exciting new automobile is bursting on the scene, at the very moment the car that would have been its direct competitor is beating an undignified retreat.

The racy little M3 from BMW and Mercedes' 190E 2.3-16 share a basic recipe: hot sixteen-valve four-cylinders stuffed into their makers' smallest sedan bodies, along with extra-firm suspensions and fast-looking aero add-ons. Both keep one eye on European Group A Touring Car racing, the other on monied enthusiasts who are powerless to resist a personal hot rod from a German factory.

Yet these two cars come across as very different animals. We now know they differ widely in terms of success on the road (we loved the sporty Mercedes, *until* we drove it back to back with the M3), and apparently they'll differ in success on the market, too: Mercedes-Benz of North America has unceremoniously dumped the 2.3-16 from its 1988 lineup.

Perhaps we shouldn't be too surprised by these divergent fortunes. Even though Bayerische Motoren Werke and Daimler-Benz drew a bead on the same target with these high-powered missiles, they were aiming from two quite different positions. Traditionally, Mercedes cars have been stout and solid, competent to a fault, almost weighed down by their own quality and excellence. Powerful engines moved them, taut suspensions kept them in control, and firm seats propped up their occupants.

By contrast, BMWs have always had a sporting flair. Smaller, lighter, more lithe, and more maneuverable, they played the wide receivers to Mercedes' power backs, the attack subs to D-B's boomers.

Is it any wonder, then, that Munich could step up to a race-based sports sedan more deftly than Sindelfingen could pare down to one? At the very least, dealer salespeople in BMW's American showrooms could speak the sporting tongue more convincingly than their counterparts at the Mercedes stores. So today we have a genuine, 192-bhp BMW street racer sure to be in hot demand, and a Mercedes "equivalent" we can only remember (or go drive in Eu-

PHOTOGRAPHY BY TOM DREW

rope, where it remains for sale).

The M3 has a clearly focused commitment to the fast-paced life. Mechanically, it blends some serious high-speed hardware with the solid 3-series base. And it comes out better—faster, to be sure, but also more responsive, better balanced, maybe even safer—than any of its 325 siblings. At $34,000, it is the most expensive of the Three family, but, remember, the 190E 2.3-16 sold (when it sold at all) for $41,150. The M3 may be the best product BMW offers in this country, especially if we give it points for continuing the marque's tradition of appealing to the sporting spirit.

Little in this world tickles such a spirit like lots of sheer power. The 2.3-liter engine was developed by Motorsport GmbH—BMW's high-output, high-profile subsidiary, and the "M" in all those exciting cars. It spins out 192 horsepower at a lusty 6750 rpm, and is almost in the modern motorcycle class for technology and tuning. If you're a fan of specific output, and of what that tells you about state of tune, here's about as close as you'll get to 100 bhp per liter without turbocharging. The M3 develops 83.5 bhp per liter. A 3.2 Ferrari makes slightly over 81; an MR2, 70; a Corvette, under 43.

This iron-block engine uses Siamesed cylinder bores to make room for big 93.4mm holes; the short stroke of only 84.0mm helps the innards to spin past 7000 rpm without trauma. Cast pistons are designed to trim weight, and they carry the wrist pin high to permit a longer connecting rod; this reduces rod angularity in changing reciprocating motion to rotation.

Double chain-driven camshafts ride in the aluminum cylinder head and operate the light 32mm and 37mm valves directly through shim-adjusted bucket tappets. Centrally located spark plugs and squish areas around the perimeters of the shallow pent-roof combustion chambers give enough turbulence and sufficiently short flame travel so that compression ratio can be all of 10.5:1. Bosch DME (Digital Motor Electronics) controls the ignition and the fuel injection. On one side of the head is a beautiful bank of four alloy intake runners, each with an individual throttle butterfly. On the other is a tubular four-into-two exhaust header.

The torque peak of 170 pounds-

Bulged fenders are subtle, sill extensions less so, high-flying rear wing not at all.

Even the M3's interior is high-performance, with BMW's great, snug-fit sport seats.

How to get 1.37 bhp from every cubic inch: lots of cams, valves, and compression.

COMPARATIVE DATA

	BMW M3	MERCEDES 190 2.3-16
Engine type	2.3-liter, 16-valve four	2.3-liter, 16-valve four
Brake horsepower	192 @ 6750 rpm	167 @ 5800 rpm
Torque	170 @ 4750 rpm	162 @ 4750 rpm
Redline	7240 rpm	6800 rpm
Compression ratio	10.5:1	9.7:1
Gear ratios	(I) 3.83 (II) 2.20 (III) 1.40 (IV) 1.00 (V) 0.81	(I) 4.08 (II) 2.52 (III) 1.77 (IV) 1.26 (V) 1.00
Final-drive ratio	4.10:1	3.25:1
Power-to-weight ratio	14.2 lb/hp	18.0 lb/hp
Wheelbase	100.9 in	104.9 in
Length	171.1 in	174.4 in
Width	66.1 in	67.2 in
Curb weight	2735 lb	3010 lb
Weight distribution	53/47%	54/46%
Steering	power-assisted rack-and-pinion	power-assisted rack-and-pinion
Brakes, front	11.0-inch vented discs	11.2-inch vented discs
rear	11.2-inch discs anti-lock system	10.2-inch discs anti-lock system
Tires	205/55VR-15 Pirelli P600	205/55VR-15 Pirelli P6

feet comes at a high 4750 rpm, further evidence of how tightly wrapped this semi-racing engine is. A five-speed gearbox with ratios that feel unusually close-spaced and short overall helps the engine run where it does its best work. In crafting the car for the U.S. market, BMW changed its shift pattern from the European sport configuration, with first outside the H, to the arrangement more familiar here, with *fifth* outside the H.

While sorting out the suspension so it could handle the new speed and power, BMW's engineers came up with a specification that brings new-found stability and dependability to the 3-series platform. True, ride height is awfully low, and spring and damper rates are quite high. But ramp clearance and ride harshness are not affected adversely enough to

make us want to give up any of the M3's fine, firm control.

Front suspension geometry incorporates much more caster to give greater directional stability. Coil-spring rates are progressive in back but remain linear in front—simply because that's what worked best. Low-pressure gas-charged shocks are incorporated in the struts in front and mounted atop the semi-trailing arms in back. The shocks fight fluid aeration and its attendant heat fade, and provide degressive damping, to give way on extreme impacts and smooth out the ride.

Ambient air gives way, too, more readily for the M3's passing than for any other 3-series, and by quite a margin. Front and rear aprons with integral bumpers, gently contoured fenders, rocker panel skirts, and a raised rear deck with a wild, angular wing all slash drag coefficient from a 325's 0.37 way down to 0.33—and that's with fattish 205/55VR-15 tires.

The rear-end treatment includes a fiberglass cover. for the C-pillars that changes the roofline subtly, reangles the backlight, and blends into the higher, flattened trunk lid. That wing is a sturdy slab constructed of polyurethane over a paper honeycomb core. Quality, fit, and match of the new pieces are flawless throughout; our only complaint centers on a spot where the rounded lower edge of the new deck lid wraps over the squared shape at the upper rear corner of the sheetmetal.

Frankly, we're also not sure whether the wing treatment isn't going a little far for a road car. It struck us as awfully boy-racerish, with that thing jutting up back there. On the other hand, the rest of the car is unobtrusive and conservative in the normal BMW way, and in any case, this look will shortly come to be widely recognized as simply the M3 style—giving it positive impact in the important circles, whether it's aesthetically perfect or not.

We may have felt more obvious on the street than we usually prefer, but the M3 left us with little else to complain about. Our time in the car began with a 600-mile waltz across New York State and southern Ontario, returning to Ann Arbor from the press introduction at lovely Lime Rock Park in Connecticut. Such a freeway run—particularly in that part of the continent, where the climate is hard on pavement—should show the

The factory-built, hot-rod sedan works much better as a BMW than as a Mercedes.

BMW M3 Power Lines

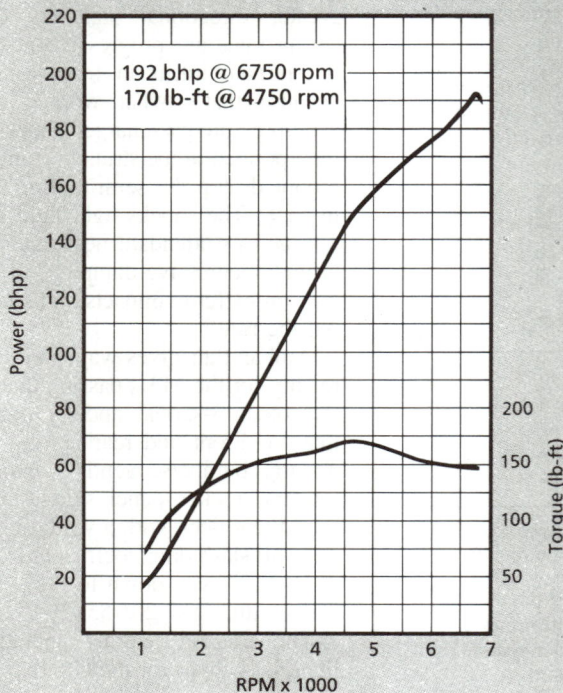

192 bhp @ 6750 rpm
170 lb-ft @ 4750 rpm

Power (bhp) — y-axis: 20, 40, 60, 80, 100, 120, 140, 160, 180, 200, 220
Torque (lb-ft) — right y-axis: 50, 100, 150, 200
RPM x 1000 — x-axis: 1, 2, 3, 4, 5, 6, 7

Mercedes 190 2.3-16 Power Lines

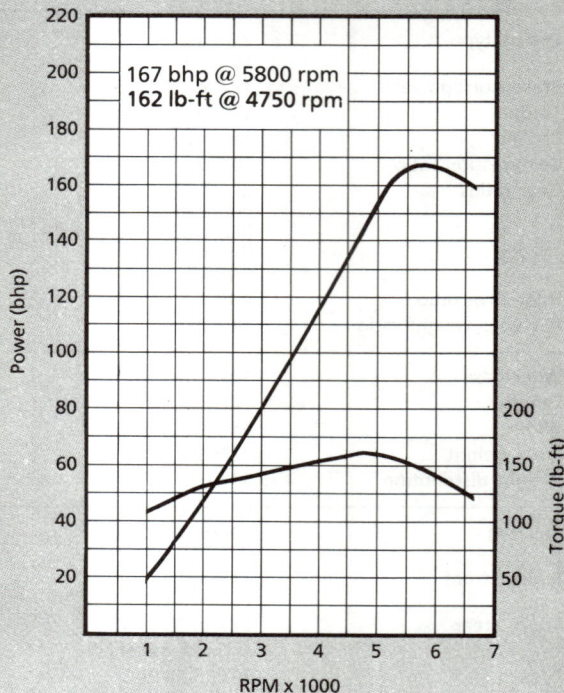

167 bhp @ 5800 rpm
162 lb-ft @ 4750 rpm

Power (bhp) — y-axis: 20, 40, 60, 80, 100, 120, 140, 160, 180, 200, 220
Torque (lb-ft) — right y-axis: 50, 100, 150, 200
RPM x 1000 — x-axis: 1, 2, 3, 4, 5, 6, 7

BMW M3

sport-tuned M3 at its worst. Well, if that's as bad as it gets, we'll take it. Yes, the car is a bit stiff-legged when hitting patches and seams at speed, but it's not too bad. Yes, its engine note is a trifle more prominent than you might prefer, but it's not too bad. Yes, engine and road vibration conspire to fuzz images in the rear-view mirror, but it's not too bad.

In short, the comfort compromises the M3 makes strike us as minor and reasonable. For a car of its type, delivering what it does in the way of performance, it doesn't really demand too much of the long-haul driver. For instance, the engine growl is obvious, but not offensive, even though it's exacerbated by the short-ish gearing (70 mph takes over 3600 rpm). Those excellent sport seats BMW uses in several of its zestier models contribute a lot, even on the Interstate. Never mind (for the moment) their aggressive side bolstering; they are perfectly shaped on the bottom as well. And the height adjustment helps place them in the car exactly the way you want them.

You'll also appreciate the M3's great seating—plus its zippy response and sharp maneuverability—around town and in the busy suburbs.

You might notice the small numb spot in the steering right around center—pretty typical of most recent BMWs we've driven—but you'll also realize that the shifter and the pedals are ideally placed, and that they work so sweetly you barely pay attention to them. Instead, you drive the car.

Of course, absolutely the best place to do *that* is out on the lightly populated back roads, preferably twisty, playful ones. Here, the M3 truly hits its favored stride. Also, here, it most clearly demonstrates its superiority over Mercedes' similar but different 190E 2.3-16. The M3 is a focused, balanced, effective sports sedan; in comparison, the 190 feels a little unsure of itself, a little confused.

Both cars are comfortable, beauti-

fully outfitted, and swift. But where the Mercedes feels soft, undertired, and resolved to plow at the nose no matter what, the BMW turns in much more crisply, gets more out of its rear tires during cornering, and is more responsive to the driver's midturn wishes. It, too, is a basic understeerer, but much less determinedly so. Its somewhat more neutral balance allows an attentive driver to obtain attitude changes (generally minor and manageable) using throttle in the bends. A quick lift lightens the tail and tightens the front; go hard back into the gas (especially over 5000 rpm), and you get a nice neutral drive out of fast turns, or a little gentle oversteer if the corner is slower and tighter.

Start working both cars hard like this, and you'll notice a couple of

PERFORMANCE*

	BMW M3	MERCEDES 190 2.3-16
0–60	7.6 seconds	8.1 seconds
Top speed	143 mph	137 mph
EPA city driving (est)	17 mpg	18 mpg

*manufacturer's data

other points the BMW scores. The Mercedes' stubby gear lever has light and lovely action, but it requires far too precise a hand—and too much attention when you're working hard—to slot into the gates correctly. The BMW's lever travels farther through a typical gearchange, but it goes where you want it without calling extra attention to itself. Also, the M3's big tachometer is much more legible, its high-bolstered seats hold more snugly, and it communicates a clearer sense of what's happening down at the tires.

Both of these sixteen-valve 2.3s are highly tuned, modest-displacement engines, both quite peaky by normal standards. They each have a little performance bubble just over 4000 rpm as they climb up onto the torque peak. The BMW's four seems to hold up a bit better off the peak, while the 190 falls off the cam into a torque pit more readily.

Looking at the torque curves for the engines confirms that impression. Although the peak values are close—within five percent, and arriving at the same engine speed—it's clear the BMW's output hangs in there better both above and below the point of maximum urge. This makes the M3 feel a little more flexible and accommodating.

As you would imagine, given the torque data and BMW's weight advantage (2735 pounds to 3010), the M3 holds the performance edge over the 190. BMW quotes a 0-to-60-mph time of 7.6 seconds versus the M-B's 8.1, and a top speed of 143, 6 mph faster than the 190. Over the broad spectrum of actual down-the-road driving conditions, however, the two cars are quite evenly matched. Credit the ideal gearing Mercedes selected for the 190, and note that the BMW's horsepower advantages are greatest above 6000 rpm. Our testing confirmed that only when both drivers work *extremely* hard does the M3 pull ahead noticeably.

Of course, in the larger race—the one in the marketplace—the BMW has pulled ahead solidly and for good. Disappointing sales and near duplication from the tamer six-cylinder 190E 2.6 finally killed the 2.3-16 in the United States. Yet there's no question that the idea of a snarky-looking factory-hot-rodded sports sedan has enduring appeal. It might not make the ideal Mercedes-Benz. But look how it works as a BMW. 🛑

1988 BMW M3

GENERAL:
Front-engine, rear-wheel-drive sedan
4-passenger, 2-door steel body
Base price/price as tested $34,000/$34,335

MAJOR EQUIPMENT:
Air conditioning standard
Sunroof standard
AM/FM/cassette standard
Leather interior standard
Cruise control standard

ENGINE:
16-valve DOHC 4-in-line, iron block, aluminum head
Bore x stroke 3.68 x 3.31 in (93.4 x 84.0mm)
Displacement 140 cu in (2302cc)
Compression ratio 10.5:1
Fuel system Bosch Motronic injection
Power SAE net 192 bhp @ 6750 rpm
Torque SAE net 170 lb-ft @ 4750 rpm
Redline 7240 rpm

DRIVETRAIN:
5-speed manual transmission
Gear ratios (I) 3.83 (II) 2.20 (III) 1.40 (IV) 1.00 (V) 0.81
Final-drive ratio 4.10:1

MEASUREMENTS:
Wheelbase 100.9 in
Track front/rear 55.6/56.4 in
Length 171.1 in
Width 66.1 in
Height 53.9 in
Curb weight 2735 lb
Weight distribution front/rear 53/47%
Fuel capacity 15.3 gal

SUSPENSION:
Independent front, with gas-pressurized damper struts, lower control arms, coil springs, anti-roll bar
Independent rear, with semi-trailing arms, coil springs, gas-pressurized shocks, anti-roll bar

STEERING:
Rack-and-pinion, power-assisted

BRAKES:
11.0-in vented discs front
11.2-in discs rear
Anti-lock system

WHEELS and TIRES:
15 x 7.0-in cast aluminum wheels
205/55VR-15 Pirelli P600 tires

PERFORMANCE (manufacturer's data):
0–60 mph in 7.6 sec
Standing ¼-mile in 15.7 sec @ 94 mph
Top speed 143 mph
EPA estimated city driving 17 mpg

MAINTENANCE:
Headlamp unit $10.62
Front quarter-panel $356.80
Brake pads front wheels $137.60
Air filter $12.08
Oil filter $3.35
Recommended oil change interval according to service indicator system

	EXCELLENT	GOOD	FAIR	POOR
ENGINE				
power	•			
response	•			
smoothness		•		
DRIVETRAIN				
shift action	•			
power delivery	•			
STEERING				
effort	•			
response		•		
feel		•		
RIDE				
general comfort		•		
roll control	•			
pitch control		•		
HANDLING				
directional stability		•		
predictability	•			
maneuverability	•			
BRAKES				
response		•		
modulation		•		
effectiveness	•			
GENERAL				
ergonomics		•		
instrumentation	•			
roominess			•	
seating comfort	•			
fit and finish	•			
storage space	•			
OVERALL				
dollar value		•		
fun to drive	•			

PHOTOGRAPHY: AARON KILEY

Dinan Turbo BMW M6

The Annihilator.

BY NICHOLAS BISSOON-DATH

• The needle races past 60 mph. The on-ramp ahead curves tightly away, but your right foot stays flat to the floor, the turbocharged engine screaming with the hard-edged bass of a highly tuned six-cylinder in full cry. As the sign suggesting a ramp speed of 25 mph flashes past, you bend into the turn, and your passenger audibly pleads for divine protection. You feel the immense lateral loadings build as your speed rises, but the car clings to the road with no dramatics. As you merge onto the highway at more than 80 mph, you look over at the disbelieving face in the right seat. Welcome to the world of the Dinan Turbo BMW M6.

AMG has the Hammer, but Dinan Engineering has an equally formidable implement; think of it as the Annihilator. It blasts from 0 to 60 mph in 4.8 seconds, through the quarter-mile in 13.2 seconds at 107 mph, and on to a rev-limited top speed of 172. It offers race-car handling, BMW quality and feel, and a shove in the back that will get anyone's attention.

Steve Dinan is a 34-year-old mechanical engineer and the founder of Dinan Engineering (81 Pioneer Way, Mountain View, California 94041; 415–962–9417). He has spent the last nine years of his life servicing, racing, and tuning BMWs. The Turbo M6 is his star, and he has every reason to be proud of its brilliance.

The source of the Dinan M6's prodi-

gious performance is its modified turbo engine. Dinan Engineering retains the 24-valve M6 block and head, but that's about all. It lowers the compression ratio·from 9.8 to 7.7:1. A specially matched Garrett T04B turbocharger supplies the boost, which is limited by a Roto-Master waste gate to 13.0 psi. A huge HKS intercooler

lowers the temperature of the intake charge; at 134 mph on a 70-degree day, according to Dinan, the reduction is a whopping 160 degrees. Dinan also installs a new airflow sensor and high-flow injectors and enlarges the stock air-cleaner inlet for better breathing. The final touch is a Bosch Motronic engine-control system, reprogrammed to Dinan's specifications by Veloz Car Computers.

The results are 390 horsepower at 6400 rpm—134 more hp than the standard M6—and a 7300-rpm redline. Not only is the power amazing, but it feeds in smoothly and progressively in response to the throttle. Press with your right foot and full boost is available just a moment later. The thrust builds at a furious pace, and before you know it you're traveling at twice your previous speed. It's easy to maintain high average speeds along remote secondary roads, picking off other cars as if multiple-jumping your way to victory in a game of checkers.

Dinan has also seen to it that the M6's superbly controllable brakes will haul you down from high speed as often as necessary. Metallic pads grab the stock rotors, and the front brakes are cooled by means of race-car-sized ducts feeding air from two gaping intakes in the front spoiler. The Turbo M6 stops from 70 mph in only 173 feet, ten feet shorter than the stock M6 can manage.

Dinan's racing experience is evident in his Stage 4 suspension. The $1938 package includes firmer shocks, stiffer springs, and adjustable anti-roll bars at both ends. Negative-camber plates in front and a special rear crossmember allow the normally fixed camber settings to be adjusted at all four wheels.

The rolling stock consists of Goodyear Eagle ZR S or Yokohama A-008R tires on BBS modular aluminum wheels. In front, 225/50ZR-16 rubber is mounted on 8.0-inch rims; the rear tires are 255/50ZR-16s on 9.0-inch wheels.

Steve Dinan personally tunes the suspensions of his cars, and he achieves impressive results. We measured 0.91 g on our skidpad; that's 0.04 g better than a Z52 Corvette with 275/40ZR-17 tires. In the real world, the Dinan Turbo M6 will scythe through a series of switchbacks at an incredible pace. Charge into a corner at more than twice the posted limit and the car holds its line precisely. Once past your apex, the tail digs in as you press hard on the throttle and unwind the steering. You can blast through corner after corner in this fashion, in an unending stream of speed and tire squeal and fury, yet remain in complete control.

At the limit the car understeers just enough to let you know that you're about to run out of grip. The rear wheels can also be provoked loose by accelerating hard in a tight corner or by sharply backing off the throttle at the limit. When the rear does let go, however, it does so slowly and predictably.

The price of this performance is a nice, round $20,000—not including the $59,000 that an M6 will cost you. For your extra twenty grand you get the rocket engine, the Stage 4 suspension, the wheels

and tires, the brake modifications, and a special Centerforced clutch.

You also get a few compromises. The Dinan M6's ride is substantially stiffer than the production car's, and the front tires tend to follow highway ruts. If you find such behavior unacceptable, Dinan Engineering offers three other suspensions for 1970-and-later BMWs, each with its own level of control and complexity. It also sells turbo kits for both 5- and 6-series BMWs. Our test car didn't have any smog controls, but Dinan builds emissions-certified turbo engines as well. We've driven Dinan BMWs equipped with both types and found that the cleansed car suffers little in feel.

Dinan's racetrack experience and development work have paid off. The Turbo M6's limits are so high, and its acceleration is so aggressive, that only the most exotic performance cars on the planet can compete with it. This is a civilized race car for the street. Drag racers, beware. And heaven help anybody who tries to keep up with the Dinan Turbo BMW M6 on a winding road. ●

Vehicle type: front-engine, rear-wheel-drive, 4-passenger, 2-door sedan

Price as tested: $80,000

Engine type: turbocharged and intercooled 6-in-line, iron block and aluminum head, Veloz/Bosch Motronic electronic engine-control system with port fuel injection

Displacement	211 cu in, 3453cc
Power (SAE net)	390 bhp @ 6400 rpm
Transmission	5-speed
Wheelbase	103.3 in
Length	193.8 in
Curb weight	3537 lb
Zero to 60 mph	4.8 sec
Zero to 100 mph	10.8 sec
Standing ¼-mile	13.2 sec @ 107 mph
Top speed	172 mph
Braking, 70–0 mph	173 ft
Roadholding, 300-ft-dia skidpad	0.91 g
Road horsepower @ 50 mph	15 hp
C/D observed fuel economy	11 mpg

Morgan+8

. . . and the sparks fly as we head down some of the country's best driving roads with Britain'
fun with the top down surely do not exist but, as Andrew Frankel reports, the most importan

meets M3

8 heirloom and Germany's £37,000 ragtop technocrat. Two more diverse ways to have
ing is what binds this rather unusual couple together. Photography by Peter Burn

Morgan hustles through well-surfaced corners at cracking pace but handling falls apart on poor B roads. Hit a bump and whole car hops sideways. Driver needs to work but rewards are ample

FORGET APPEARances, the vast price difference and the chalk and cheese chasm in technical detail: Britain's belt and braces Morgan +8 and Germany's computer literate BMW M3 convertible do the same job. And you know it the first time you drop their tops and aim them down a favourite road on a sunny morning. It's not just that either blasts to 60 in six wind-in-the-hair seconds or less, it's that both offer an unforgettable drive. But what conclusively different cars they are.

This is indeed a prizefight with a difference: the Morgan, hairy-chested and with the support of the crowd, facing a car trained by science and the racetrack to the peak of physical fitness. The battleground: the Lake District and the Yorkshire Moors . . . and the sort of roads for which both cars were born.

In the whole of motoring there are surely not two more opposed routes to what is ultimately the same result. The Morgan gets its punch from the old and ubiquitous pushrod Rover 3528cc V8. It produces 190bhp at 5300rpm and 220lb ft of torque at 4000rpm. When this kind of power is dropped into a car weighing little more than 2000lb, shattering performance is guaranteed.

With sliding pillar front suspension and a live rear axle located by leaf springs — ye gods, even lever arm dampers — the Morgan's suspension is pure pre-war. The car costs £17,703 and you will have to wait up to four years for delivery.

The BMW's 2302cc four-cylinder engine has twice as many valves per cylinder, operated by two overhead camshafts. The engine is unique to the M3 and was designed primarily as a racing unit. In roadgoing tune it develops 200bhp at 6750rpm and 177lb ft torque at 4750rpm.

Suspension sounds identical to that of any other 3-Series: MacPherson struts up front, semi-trailing arms at the back, coil springs and telescopic dampers all round. But BMW Motorsport has been to work here. With a combination of fine tuning and geometry revisions it has produced a chassis of rare quality.

BMW GB charges a ridiculous £37,250 for the M3 convertible. Only 40 left-hand-drive cars are being imported this year.

Straight-line performance in either car is sensational but the Morgan is the quicker. It can reach 60mph in just 5.6secs and up to 100mph in 16.4secs. The BMW manages 6.0secs and 16.6secs. Above 120mph the German's vastly superior aerodynamics mean the BMW can start to regain lost ground and pull away from the Morgan. In the gears, too, the Morgan is quicker. Between 50-70mph in fourth the car accelerates in 5.1secs, and in fifth 7.6secs; the BMW requires 6.4 and 9.5secs. It is only on top speed that the Morgan has to give best. Its 122mph cannot live with the BMW's 144mph.

Driving the cars on public roads shows this performance gap to be even wider than the figures suggest. For much of our two-day jaunt the Morgan was simply held up by the BMW, at least on straight roads. One event illustrates the point: overtaking a column of dawdling traffic, I dropped the BMW down from fourth to third and powered past on full throttle. The Morgan came too, but requiring neither a gearchange nor a foot on the floor.

It may seem hard to believe that Rover's ageing pushrod V8 could be a more effective engine than the multi-valve marvel produced by BMW Motorsport, but in these cars there is no contest. First there is the Morgan's torque. Even with very high gearing — 27.6mph/1000rpm in top — it will pull cleanly from walking pace in any gear. It will continue to deliver a flood of power to the 5500rpm limit we imposed, with no 'coming onto the cam' or fall off in power at high revs — it delivers its performance in one clean, solid shove. Then there is the noise. A classic V8 burble at low revs, rising to a deep-chested roar as it is extended. The effect is inspiring.

Taken in isolation, the BMW four is a great engine. It has a tremendously successful track record and in road trim it combines its power with a quality edge that the V8 lacks. But still there are racing traits: below 4000rpm, for instance, there is not much urge. Above this it unleashes its power and the rev-counter will charge to the limiter at 7400rpm with superb response. So you have to work at the BMW, with frequent gearchanges the key to keeping it on the boil. Fail to do this and the power disappears from under your feet.

There is not the engine music either. The engine sounds less distinguished than that of a Golf GTI up to 5000rpm and there's some roughness too — certainly it's not as smooth as any BMW six-cylinder unit. After 5000rpm it issues a mechanical howl to remind you of its racing aspirations, but still it does not seduce like the Morgan.

The BMW waits until the corners before it seduces you. The M3 has as good a claim as any to having the most competent *and* entertaining front-engined chassis in production. Remarkably, the convertible has lost none of the saloon's ability. Turn-in is sharp and the grip from the 225-section Michelin MXX tyres is of the very highest order.

The convertible will understeer or oversteer on demand, but its basic cornering stance is one of strong neutrality. Push on harder and it just feels better and better, neutral cornering ▶

M3 convertible has lost none of saloon's ability. Chassis is one of the best. Steering full of feel, turn-in sharp, and car will understeer or oversteer on demand. Grip is of the highest order

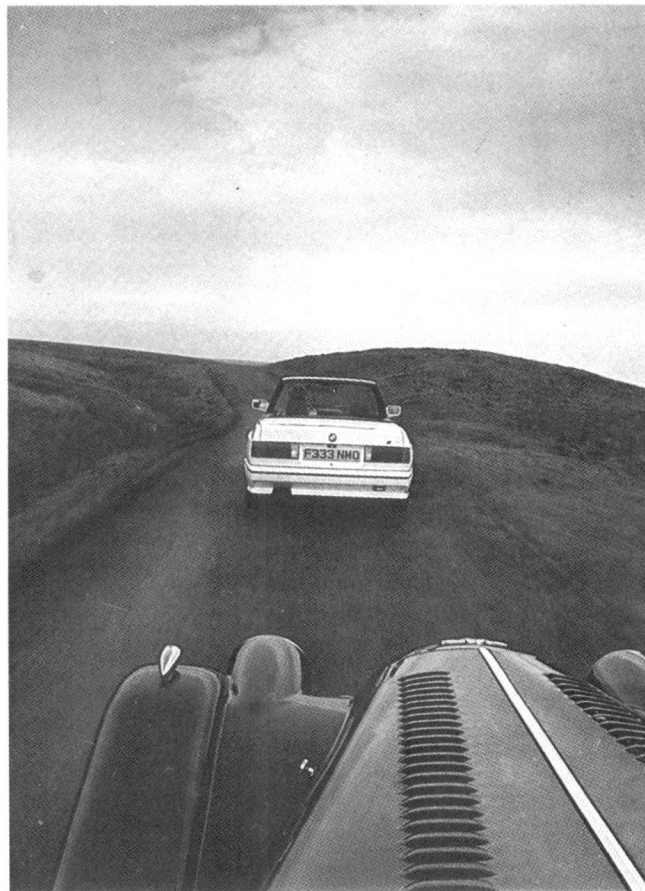

	BMW M3 Convertible	Morgan +8
DRIVETRAIN		
Cylinders	6, in line	8, 90 deg V
Capacity	2302cc	3528cc
Bore/Stroke mm	94/84	89/71
Valves per cyl	4	2
Valve operation	dohc	sohc
Compression ratio	10.5:1	9.8:1
Induction	Bosch ML-Motronic fuel injection	Lucas 'L' fuel injection
Power/rpm	200bhp/6750	190bhp/5300
Torque/rpm	171lb ft/4750	220lb ft/4000
Gearbox	5-speed manual	5-speed manual
Drive	Rear wheel drive	Rear wheel drive
Final drive ratio	3.25:1	3.31:1
Mph/1000rpm top	21.3	27.6
Tyres	Michelin MXX 225/45 ZR16	Uniroyal 205/60 VR15
DIMENSIONS		
Length (ins)	171	156
Width (ins)	66	63
Height (ins)	54	52
Wheelbase (ins)	101	98
Track f/r (ins)	56/56	53/54
Kerbweight (lb)	3105	2022
Distribution f/r	50/50	48/52
PRICES		
Total in GB	£37,250	£17,703
FUEL CONSUMPTION		
Overall mpg	22.2	20.9
Fuel tank (gals)	12.2	14
TOP SPEED		
Mean	144	122
Best	146	126

ACCELERATION (secs)		
0-30mph	2.1	2.0
0-40	2.9	2.9
0-50	4.7	4.3
0-60	6.0	5.6
0-70	8.0	7.9
0-80	10.7	9.7
0-90	12.0	12.3
0-100	16.6	16.4
0-110	22.5	21.3
0-120	28.3	29.6
Standing ¼ mile	15.8secs/90	14.4secs/95
Standing km	27.7secs/119	26.7secs/117

In each gear mph	top	4th	3rd	2nd
BMW				
10-30	—	8.7	5.6	3.5
20-40	10.6	7.3	4.6	3.1
30-50	9.2	6.5	4.2	2.8
40-60	9.3	6.4	3.9	2.9
50-70	9.5	6.0	3.9	—
60-80	9.2	6.0	4.3	—
70-90	9.5	6.1	—	—
80-100	10.0	7.6	—	—
90-110	11.2	9.0	—	—
100-120	15.2	10.2	—	—
MORGAN				
10-30	8.6	5.9	4.0	2.3
20-40	7.2	4.9	3.4	2.2
30-50	6.8	4.6	3.2	2.3
40-60	7.3	4.8	3.3	2.5
50-70	7.6	5.1	3.4	—
60-80	8.1	5.2	3.6	—
70-90	9.4	5.6	4.7	—
80-100	11.7	6.4	—	—
90-110	15.7	8.5	—	—

Both cabins well finished and comfortable, Morgan's seats compensating in part for ride deficiencies. Refined interior of M3 is further improved by supportive and heated leather seats

◀ balance eventually giving way to mild, benign oversteer. The steering, so full of feel, lets you keep the front wheels pointing in the desired direction, without drastic correction, and the car follows this line faithfully. It has no hidden vices, no ghastly secrets.

Driving the same road in the Morgan induces acute culture shock. Grip is not the problem if the road is smooth. With only 2000lb to persuade to change direction, the 205-section Uniroyals allow the +8 to be hustled through well-surfaced corners at a cracking pace. Put it on the pockmarked B roads of the Lake District and the story is very different. The car hops wholesale across the road as soon as look at a bump. The ride is truly appalling.

And the otherwise dead steering can generate the sort of kickback that wrenches the wheel from your hands. You drive this car from the seat of your pants. Do this, and it is not without its rewards. Fight the steering, kill the heavy understeer with a bootful of throttle, be ready to catch the inevitable tail slide and you will have one of the most invigorating rides this side of a rollercoaster. Despite the dead steering, the Morgan can be placed accurately, but it takes practice.

On the practical side, the BMW is streets ahead, and Morgan wouldn't have it any other way. Hood up in the M3 you could be in a saloon up to about 70mph. Engine noise is prominent, and there is a distant rumble from the fat Michelins, but wind roar, though audible at motorway speeds, is very well suppressed.

The refinement is heightened by heated leather seats which are comfortable and supportive. With comforts like these the worries of left-hand drive soon disappear.

Driving the Morgan roof-up on the motorway is not recommended. The tall gearing keeps engine noise to a minimum but since the wind drowns any attempt to hear anything, it's rather academic. The wind causes the hood to billow skywards — creating some much needed headroom — and assaulting you from every hole in the ill-fitting side-screens.

The Morgan does do some practical jobs surprisingly well. The seats are comfortable even if the ride is not. The two-stage heater keeps you warm in freezing conditions. The driving position is not terrible, even if it is short on leg room. However, none of this can make the car anything but fatiguing to drive in less than ideal conditions.

The M3 convertible is the only BMW to come with an electric hood. Raising or lowering it is a simple matter of pulling two levers and pressing a button. The mechanism required to achieve a taut cover that is both water and air tight is not to be underestimated. The hood has to go through a complex range of manoeuvres, all of which are achieved with

absolute millimetre-perfect precision.

The Morgan has a typically belt and braces hood. A skeleton frame provides the basic shape. You have to hang the hood over it and clip it to the top of the windscreen and the back of the car, having already screwed on the sidescreens. The job could conceivably be done inside five minutes. Unfortunately our time with the car was spent in freezing conditions with a howling gale, when the job becomes nigh on impossible.

Both cars are beautifully built. Scuttle-shake, which can reduce a sound saloon into a rattling undesirable, is only apparent in the BMW on badly broken surfaces. Paintwork is deep and lustrous, and body panels fit tightly and evenly.

If anything, the Morgan is more impressive. The test car was Morgan's demonstrator and even after 40,000 miles of the suspension trying to shake the car to pieces it still felt and looked like new, save for the odd stone chip. Drive one and you will know that this is no mean achievement.

The BMW teaches lessons you never forget. It extends the boundaries which the comprised structure of a convertible has previously had to observe. It is fast and flattering, and as practical a drop top as anyone could wish.

Still, there is something not quite right about this car. It's partly in the price. A 325i convertible costs £17,000 less. A saloon M3 is nearly £14,000 cheaper. This cannot be justified by leather seats and an electric hood. The car seems to have been conceived as a money-making exercise.

The Morgan has no such problems. Apart from straight-line speed, it is no match for the BMW and nor, perhaps, would it want to be. What it offers is an unrivalled tactile experience. You can get out of the BMW, unruffled, after a hard blast down a fell road and marvel at the car's ability. Do the same in the Morgan and you get out with a real sense of achievement.

Then there is the way the car looks. Beside the Morgan the BMW, for all its flared arches and spoilers, looks anonymous. The Morgan looks classically beautiful. It has a hint of fragility that makes you want to look after it. For all the money it costs, the BMW is much less of an individual.

The Morgan has only one real problem. It is pointless driving it in anything other than ideal conditions. The car's comprehensive inability to transport its occupants for long distances in anything but severe discomfort is something that only the most die-hard nut will discount. But when the roads are dry and the sun shines, you cannot have too much of it. The Morgan ladles out fun like the BMW never could.

The essence of it is that in the BMW you enjoy the car, while in the Morgan you enjoy yourself. ∎

M3 is only BMW with electric hood. Raising or lowering it is simplicity itself. Morgan's hood posed problems for testers in howling gale but whole operation can take less than five minutes in ideal conditions

BMW TOURING

M3 & M 635 CSi

Two Bavarians challenge the *Alpenstrasse*

BY JOHN LAMM
PHOTOS BY THE AUTHOR

▽

THE ALPINE ROAD is legendary for its beautiful views and challenging driving opportunities. The *Deutsche Alpenstrasse* runs east from Lindau, West Germany, at the northeast edge of the Bodensee, for 295 miles to Brechtesgaden, near Germany's eastern border with Austria. And a great run it is.   Friends and I took a pair of BMW M cars for the eastern one-third of the *Alpenstrasse*. The M6 or, as some call it, the M 635 CSi, is the senior member of that line and our M3 the one most often seen in the U.S.   We picked up the cars in Munich and took the number 8 *Autobahn* southeast toward Salzburg. About 30 km from Munich's center we turned south, heading for Tegernsee. It was Sunday and as good an excuse as any to stop in Tegernsee and sample strudel by the lake, washed down with a cup of strong coffee.   The drive east from Tegernsee to Schliersee and on to Bayrischzell is a good warm-up for the good driving that comes later. You could also have great fun by temporarily leaving the

spiel voran:
ator-Technik.

A 8 München
Salzburg

Traunstein   17 km →
Siegsdorf   11 km
Ruhpolding   3 km

305   Reit im Winkl   21 km
← Deutsche Alpenstraße

Biathlon Bundesleistungszentrum

Alpenstrasse and going instead by way of Austria, via the Urspringpass, Kufstein and Walchsee. It is as much fun as it is beautiful.

Our *Alpenstrasse* route gave us a variety in climates on our late spring drive. It was 74 de-

grees and short-sleeve weather just off the *Autobahn*, but we had to put on our ski jackets about two valleys to the south.

Reit im Winkl is the halfway point on our abbreviated journey, a small, vest-pocket town set at the base of a mountain in the Chiemgauer group. We stayed at the Hotel Unterwirt, a typical example of the nicely done German guest houses, finished in light wood and white plaster. The rooms are very comfortably furnished and the beds covered with down comforters. The hotel's restaurant is excellent, the food plentiful and the Pils perfect.

We found there is still something very special about the BMW 6-Series coupe. I'm not the only person who feels that way, obviously, because there are many new models influenced by the coupe's shape. The Acura Legend coupe and the new Ford Thunderbird come to mind immediately.

The coupe reminds me of the best of aristocratic Europe. Say of the gentleman who arrives at dinner in mid-February. He's wearing a classic pinstripe suit that is beautifully tailored. He's well tanned after a week's skiing at St Moritz. He gracefully mastered every slope by day; held court in the finest restaurant at night.

In contrast, the BMW M3 reminds me of a hard-charging young executive. He rushes into dinner dressed in whatever Hugo Boss has just made fashionable. This man is also tanned from skiing, but he's just in from Aspen. He attacked the deep powder every day; up late each night trying another trendy restaurant.

Both cars have the same fundamentals. The interiors are basically black with easy-to-read instrument panels that tell you everything you would care to know about the car's systems. BMW does this as well as any automaker in the world. The seats in both cars are excellent, but quite different. They are firm (these are, after all, German cars); the M3's seats hold you in place in street-racer fashion, but the M6's seats are broader and more luxurious.

Fit and finish are superb, and what the BMW interiors may lack in warmth, they make up for with unerring purpose: to go quickly.

The engine basics are also the same. Twin-cam, 4-valve technology is so popular and highly touted these days yet many automakers find it old hat. The M1 engine had its 24-valve head in 1978, when it bumped up horsepower from the standard 3.5-liter's 218 bhp to 277 bhp at 6500 rpm. These days the same engine, in the emissions trim the Germans and Swiss require, has 260 bhp, although there is a non-catalyst version that produces 286 bhp. U.S. editions have 256 bhp at 6500 rpm and 212 lb-ft at 4500 rpm, so Americans no longer take a real bite when it comes to performance.

That same 4-valve thinking has been applied to the M3 4-cylinder, and its 16-valve version has 192 bhp in the U.S. (195 bhp in Germany) with a catalyst and 200 bhp without it. That's very impressive: 83 bhp per liter in American form. Torque for us is 170 lb-ft, which is only a few less than the few non-catalyst M3s that are built. And both engines look the part, with BMW M POWER shining from the cam covers, and a neat row of injection horns. You would be proud to show off the engines of this pair even if they were slow, which they are not.

Their performance befits the motorsports heritage. BMW claims a 0–60 mph time of 6.8 seconds for the M6, and 7.6 for the M3. How-ever, Road & Track recorded a 0–60 mph time of 7.1 sec for a U.S. version M3 in its February 1988 road test.

Because they are at home on the *Autobahn*, both cars have impressive top speeds. For American models, BMW claims a top speed of 143 mph for the M3 and 150 mph for the M6.

The people on our *Alpenstrasse* trip chose the M 635 CSi as the best *Autobahn* cruiser. With a route that combined both the high-speed highways and smaller secondary roads, the big coupe's ride and general refinement made it the choice. There is less general busy-ness in the car with its longer gearing and bet-ter noise insulation.

Given the tight roads of the *Alpenstrasse*, the M3 won. There's certainly nothing clumsy about an M6, of course, but the size of the M3, along with the firmer suspension and fat tires,

These are the things which make European trips memorable: Old World markets such as those in Salzburg, fine wines at the Hotel Unter-wirt and breathtaking landscapes like the Chiemsee.

made it a natural to win on the twisties. It has initial understeer, of course, but with plenty of punch and good manners to do your bidding. This same firmness, of course, gave the *Autobahn* nod to the M6.

From Reit im Winkl we continued east for Berchtesgaden. This section is perhaps the most beautiful section of the *Alpenstrasse*. Whether the Schwarzbachwacht Pass or the Weissbach Gorges, the road is challenging when the traffic is light. There are plenty of restaurants along the way, but I would suggest a picnic lunch. If you have an M3 you can use the rear spoiler as a picnic table.

From Brechtesgaden you can backtrack a bit to get to the *Autobahn* via Bad Reichenhall. But at this point it is only minutes to Salzburg. There is Hohensalzburg castle that dominates the city and Mozart's birthplace, but we opted for Tomasellis, said to have been Mozart's favorite pastry shop. We ordered the variety tray and coffee *mit Schlag*. We wandered back to the cars through the open street market shops that were selling fruit, nuts and flowers.

On the way back to Munich via the *Autobahn*, we turned off for a short excursion to Prien, on the Chiemsee, the largest lake in Bavaria. The view across the lake to the Herrninsel island with its beautiful King Ludwig II castle, set off by the Alps behind it, is very much worth the stop.

Then we motored back to Munich, heading for the BMW enclave with its distinctive tall columns that can be seen from a great distance. Being the easiest part of the trip, it is also where we finally got lost.

M3 U.S. SPECIFICATIONS

Price	$34,950
Curb weight, lb	2865
Wheelbase, in.	101.0
Track, front/rear	55.6/56.1
Length	171.1
Width	66.1
Height	53.9
Fuel capacity, U.S. gal.	14.5

ENGINE

Type	dohc 4-valve inline-4
Bore x stroke, in./mm	3.68 x 3.31/93.4 x 84.0
Displacement, cu in./cc	140/2302
Compression ratio	10.5:1
Bhp @ rpm, SAE net	192 @ 6750
Torque @ rpm, lb-ft	170 @ 4750
Fuel injection	Bosch Motronic

DRIVETRAIN

Transmission	5-sp manual
Gear ratios: 5th (0.81)	3.32:1
4th (1.00)	4.10:1
3rd (1.40)	5.74:1
2nd (2.20)	9.02:1
1st (3.83)	15.70:1
Final-drive ratio	4.10:1

CHASSIS & BODY

Layout	front engine/rear drive
Brake system, front/rear	11.0-in. vented discs/11.1-in. vented discs, vacuum assist, ABS
Wheels	cast alloy, 15 x 7J
Tires	Pirelli P600, 205/55VR-16
Steering type	recirculating ball, power assist
Turns, lock-to-lock	3.9
Suspension, front/rear: MacPherson struts, lower A-arms, coil springs, tube shocks, anti-roll bar/semi-trailing arms, coil springs, tube shocks, anti-roll bar	

PERFORMANCE

0–60 mph, sec	7.1
Standing ¼ mile, sec @ mph	15.4 @ 91.0
Top speed, mph[1]	143

[1]Factory claim

M 635 CSi U.S. SPECIFICATIONS

Price	$55,950
Curb weight, lb	3570
Wheelbase, in.	103.3
Track, front/rear	56.3/57.7
Length	193.8
Width	67.9
Height	53.3
Fuel capacity, U.S. gal.	16.6

ENGINE

Type	dohc 4-valve inline-6
Bore x stroke, in./mm	3.68 x 3.31/93.4 x 84.0
Displacement, cu in./cc	211/3453
Compression ratio	9.8:1
Bhp @ rpm, SAE net	256 @ 6500
Torque @ rpm, lb-ft	243 @ 4500
Fuel injection	Bosch Motronic

DRIVETRAIN

Transmission	5-sp manual
Gear ratios: 5th (0.81)	3.17:1
4th (1.00)	3.91:1
3rd (1.35)	5.28:1
2nd (2.08)	8.13:1
1st (3.51)	13.72:1
Final-drive ratio	3.91:1

CHASSIS & BODY

Layout	front engine/rear drive
Brake system, front/rear	11.8-in. vented discs/11.2-in. discs, hydraulic assist, ABS
Wheels	forged alloy, 415 x 195TR
Tires	Michelin TRX, 240/45VR-415
Steering type	recirculating ball, power assist
Turns, lock-to-lock	3.5
Suspension, front/rear: MacPherson struts, lower A-arms, coil springs, tube shocks, anti-roll bar/semi-trailing arms, coil springs, tube shocks, anti-roll bar	

PERFORMANCE[1]

0–60 mph, sec	6.8
Standing ¼ mile, sec @ mph	na
Top speed, mph	150

[1]Factory claims   na means information not available

What's in an M?

IT'S USUALLY QUITE subtle, just a stylized M with the well known red, violet and blue BMW colors on the trunk lid and grille. But beware, because even though the car may look stock, and any changes might appear to be as subtle as the badge, this car has been gone over by the men at Motorsports. It is no longer a normal Bimmer.

BMW Motorsports was formed in 1972 to give the company an official sports department. Initially that meant racing. Remember the high-winged BMW coupes, and the company's Formula 2 and Junior Teams, with Eddie Cheever a major player in both? The first step into production by the "M group" came in 1978, when the M1 mid-engine Grand Touring car was designed and developed by Motorsports, Giugiaro, Lamborghini and Bauer.

When that car's run was through, it seemed a shame not to continue to use the twincam 24-valve version of the company's highly regarded in-line-6 that was done for the M1. And with this new need came a new direction for Motorsports: to keep BMW's hand in racing, and to become more production-car oriented.

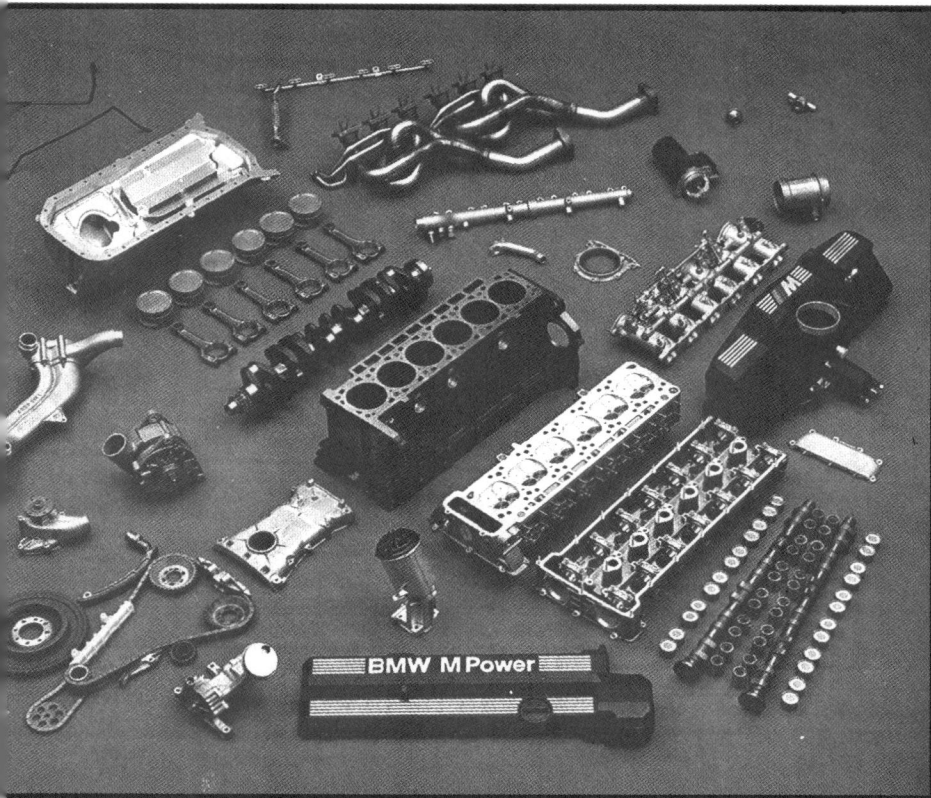

The M1 engine was tried in the 5-Series sedan, but the engine's first series production was the 6-Series coupe. Horsepower was increased from the stock 635 CSi's 218 to the current version's 260 in catalyst form or 286 without one. There were appropriate changes to the suspension, etc, and the result, called the M 635 CSi, is only now going out of production. The second series M car was the original M5, with the M3 making its debut at the Frankfurt Show in 1985. Newest of the line is the recently introduced M variation of the new 5-Series sedan.

In fact, Motorsports doesn't really build any automobiles. Based in a small building a short drive from BMW headquarters in Munich, the sporting side of the company designs and develops the M series cars, which are then assembled on the production line with the standard versions. This same basic no-build approach was used when BMW was in Formula 1 racing, with Motorsports supplying engines to outside race teams. In its recent successful forays into touring car racing using the M3, Motorsports has built only the race cars it needs to develop the product. Outside teams then buy their racers as gigantic kits: finished engine, body-in-white car with the rollcage installed, a whole bunch of parts and a book that explains how to put it all together.

But that's only a small part of the company's annual income of some $20 million. BMW M Technic creates accessories for the company's customers, ranging from steering wheels to road wheels to special aerodynamic body kits. And there are the luxury items, like a full leather interior for your 6-Series coupe, including a trunk lined with the same material. Like Porsche's Sonderwunch program, Motorsports will do whatever you'd like for your BMW as long as it isn't illegal or immoral. There is also a line of clothing with the M logo.

Like Porsche, this personal tailoring approach also gives BMW a chance to compete with the German specialty tuners, like Alpina or Koenig. Motorsports' advantage over the small firms is the resources of the big automaker. M3s and M5s have been developed with the company's usual thoroughness, such as 25,000 miles of high-speed testing at Nardo in Italy and more than 6000 miles of durability laps on the old Nürburgring course. There's the usual bench testing of the engines and the quality controls of the assembly line. Tack on a normal factory warranty. And go like hell.

That's why you should show a little extra respect for the BMWs with the subtle little M on the decklid and grille.—*John Lamm*

Schizophrenic style

The BMW M3 Convertible is a contradiction in terms. On the one hand there are the homologation special mechanicals from the M3, implying peaky performance and lack of refinement. On the other, the 325's fold-down top and luxurious interior, suggesting the soft life. It could be either a brilliant combination or wildly schizophrenic…

The exploits of Messrs Sytner, Weaver *et al* on the tracks have proved that those M3 mechanicals can deliver the goods performance-wise. Under the bonnet there's that unique 2302cc in-line four, with four valves per cylinder and twin camshafts which, even in road trim, produces an impressive 200bhp at 6750rpm, and 177lb ft of torque at 4750rpm. It's coupled to a Getrag five-speed gearbox with racing change pattern – first is a dog-leg left.

Suspension is 3-series derived, with MacPherson struts up front and semi-trailing arms at the rear, plus discs all round, but of course BMW Motorsport's efforts in fine-tuning and geometry alterations, helped by fat Michelin MXX 225/45ZR16 tyres, ensures that the chassis can cope. The M3 Convertible is based on the original M3, not the evolution model, by the way.

The rest of the car is a combination of 325i Convertible with some M3 panelwork, such as the wheel arches, to accommodate those wide tyres. The weight loss resulting from the removal of the metal roof is more than made up by the extra stiffening of the chassis and the roof mechanism; the Convertible weighs some 350lbs more than the standard M3. It is too a true convertible, in that the top disappears when down, and there's no roll-over bar.

This hasn't had a great deal of effect on the performance. The maximum speed is 144mph (top up, of course), and the 0-60mph time is 6 seconds dead, which is exceedingly rapid by anybody's terms. The engine is a willing charger, depending on revs, revs and more revs to deliver the goods, which takes care of the performance side of the car's character; on the hedonistic side, there's not a great deal of urge below about 4000rpm, and as the 7000 red-line is approached its racing heritage comes out in some raucousness and harshness. BMW has done a pretty good job in refining it, but then if you want refinement you'd presumably go for the standard 325i drop-head with its silky smooth 6-cylinder power unit.

The race-perfected suspension means that the Convertible is every bit as brilliant in the corners as the standard M3. If you need a word to sum it up, it would be 'bite': it corners dead flat, very neutrally, and with immense confidence. Steering precision and weighting is absolutely ideal, too. As

When is an M3 not an M3? When it's an M3 Convertible… The removal of the roof and the fitment of a soft-top transforms the car (above), making it ultra-desirable to many, but overpriced to an equal number. For all the alterations, the 2.3-litre engine (right) remains a jewel.

an utterly exhilarating cross-country machine, on a bright sunny day, the M3 Convertible is near perfection.

Assuming our sybaritic owner lives in a city, however, he might find life a little less pleasant at times. For a start, the Convertible is only available in left-hand drive form which, with those invisible flared arches, makes placing the car in traffic a more cautious operation than usual; secondly there's that dog-leg first gear

The M3 Convertible is an instant collectors' item

change. Not only is it an unusual pattern but that on the test car I'd want it improved dramatically so as not to wrong slot at quite the wrong time. On top of that, at low speeds the brakes are over-servoed. All in all, the Convertible is not at its best as a pose machine in London's West End.

On the other hand, our hedonist would have no complaints about the interior or the roof. The leather seats are properly supportive, if a little compact in the back (the top has to go somewhere), and the instruments and controls pure BMW, which means excellent. Top up, you have to be cruising at the ton or more before wind thrash on the roof drowns out normal conversation. Top down and with side windows raised there is very little buffeting until you are, again, exceeding the speed limit, and even then it's not excessive. As has been said, on a clear day and with the roof down, it's everything a sun-worshipping enthusiast could want.

And if you really want to impress the neighbours, operate the roof when you know they're watching. All you do is unclip a couple of levers on the top of windscreen, then press a button. There's a whirring and a buzzing as the rear flap opens, the roof sails majestically up then down into its bay, and the flap closes – all automatically. It's guaranteed to leave them goggle-eyed.

So: is the convertible a brilliant combination, or merely schizophrenic? The price – a whopping £37,250 – suggests that BMW think the former: in practice there's a tendency towards the latter, which, it must be said, adds to the character of the car. Dull it is not. Perfect it is not either. One thing is for sure, though: the M3 Convertible is an instant collector's item. In years to come it will be seen as an investment. Is it therefore overpriced? Who knows? ∎

BMW M3 Convertible
£37,250

Cylinder/capacity	4 in-line, 2302cc
Bore/stroke	94/84mm
Valve gear	Twin ohc, 4 valves/cyl
Fuel system	Bosch ML-Motronic
Power/rpm	200bhp/6750rpm
Torque/rpm	177lb ft/4750rpm
Maximum speed	144mph
0-60mph	6.0s

BMW M5

The temptation is to slow down more than you have to just to blip the throttle and revel in the glorious noise as you accelerate back up to speed/Art Markus

BMW's new M5 is far more svelte than the previous model, which is in keeping with the very first Motorsport model, the glorious M1

When a lucky buyer orders a BMW M5, he is quite likely to be invited to Munich to see the M5 – possibly even his own car – taking shape on the special production line at BMW Motorsport.

You can just imagine it, can't you. It's early evening in some trendy wine bar in the city . . . or possibly Sunday lunchtime in a cosy pub in the Surrey stockbroker belt. As so often happens when a group of men get together, the conversation turns to cars. One-upmanship is the name of the game. But there are rules, and the cardinal rule is that you mustn't look as if you're trying. Indeed, ideally you shouldn't even look as if you're *playing*.

"I've just been to see my new M5 being built in Munich," one says, as casually as he dares. (Our friend doesn't feel the need to say it's a BMW. Most of his pals are interested in cars anyway – after all, they all drive Porsches, or BMWs, or Audis, or Mercedes – and they know that M5 means BMW. Anyhow, he's been talking about it for months.)

"Well of course the MD of the dealership invited me personally, you know." (Here, our friend is emphasising just how important a customer he is – you know, me and the MD, we're like *that* – and, indirectly, what an expensive car his M5 is.)

Of course, he's probably right: any M5 buyer *is* an important customer. Some of his pals grate their teeth – our friend is scoring lots of points here.

"They build them separately to the other cars, you know," (this of course implies some innate superiority over those nasty mass-produced Porsches, Audis, and Mercedes . . . not to mention lesser species of BMW!) "at Motorsport." (Again, no need to mention the name – they all know he means BMW Motorsport. And dropping the word Motorsport into the conversation implies that his M5 is being built alongside those gaudily-liveried M3 touring cars they've all seen on Saturday afternoons on the telly. It's not, in fact – *Rennabteilung* is completely separate, although it is under the same roof – but they don't need to know that.)

"They are hand built, you know." (Just rubbing in the exclusivity a little bit here.) But that is not strictly true. Hand assembled would probably be a more accurate description. The 535i bodyshells arrive more-or-less complete, having been taken from the normal 5-series production line. And most of the other mechanical components arrive at Motorsport as subassemblies. Indeed, a whole row of those mighty 24-valve, six-cylinder engines waiting to be installed, complete with gearbox and front suspension, makes an awesome sight.

But it is true that the major mechanical assembly is handled by just one technician or, if they prefer to work as a team, perhaps two. All 72 of the technicians involved in assembly of M5s at BMW Motorsport are highly trained, highly qualified and highly skilled. Time was when the term hand built was synonymous with only the very highest standards of assembly. Nowadays, it is generally acknowledged that modern sophisticated production line methods of motor vehicle assembly are probably at least as good as, if not even better than, can be achieved by even the most skilled craftsmen.

What is an advantage, though, is the great flexibility that the Motorsport operation gives BMW. For instance, the M5 can be ordered with almost any permutation of options. As an example, the rear seats can be either individual sports seats, which of course makes the car strictly a four-seater, or the more conventional rear bench seat, which makes the car a full five-seater. If the customer *really* has money to burn, he can even specify a leather-trimmed boot!

Each car takes about a week to complete, and before it is released, it is subjected to the most thorough and painstaking inspection, taking four or five hours, including a test drive. The inspection is so thorough, in fact, that the inspectors reckon, with a bit of practise, to be able to recognise from certain telltale signs, who has assembled each particular car.

A visit to BMW Motorsport in Garching is enough to impress even the most cynical of motoring journalists. But for BMW, it is a powerful marketing tool. For most customers, however blase, it must be quite exciting to be invited to BMW Motorsport, and to watch their own car, or others just like it, taking shape. And flattering to be asked . . . .

But for any potential buyer who may be wavering, a visit to BMW Motorsport is sure to clinch the deal. No one with an ounce of enthusiasm for the motor vehicle could fail to be moved by the sight of painted but stripped bodyshells arriving at Motorsport on trolleys; or stacks and stacks of crates containing complete sub-assemblies – front suspensions, rear axles, and so on; or rows of engines waiting to be installed. And if you haven't got that necessary ounce of enthusiasm, please don't buy an M5 – you don't deserve one.

This, surely, is one of the finest automobiles ever made. Despite the fact that it is now equipped with a three-way catalytic convertor, the 24-valve, six-cylinder engine is even more powerful than before, up to 315bhp thanks in part to a small capacity increase. It is smooth, powerful, responsive . . . and makes the finest noise I know. There is a great temptation to slow down more than you have to, and to change gear frequently, just to blip the throttle on downchanges, and then revel in the glorious noise as you accelerate back up to speed. Of course, if you ever got tired of that (you'd have to be brain dead!) the engine is so tractable you scarcely have to change gear at all. What a jewel!

And that gorgeous engine is matched by the rest of the car. It goes without saying that the M5 has all the comfort and safety features that buyers in this category expect. It is very fast; the engine is electronically governed to give a top speed of 155mph, with acceleration to match – 0-60mph in just over six seconds.

The brakes are superb, the handling and roadholding the same. I wouldn't change a thing. Really, the only thing you could criticise is the price – I can't afford one! At £43,465, the M5 is obviously destined for only a lucky few. I would imagine that every time you climbed in and started that magnificent engine, it would seem worth every penny. . . .■

HALF THE trouble in assessing cars like the BMW M5 is to find an appropriate pigeonhole in which to put them. With what do you compare the latest M5, in order to decide whether it's worth the requisite 43 and a half grand? Is it, for instance, dead of steering, sticky of brake and ponderous of body when compared to a Lotus Esprit, which costs a similar amount? Ridiculous – the BMW has more seats, more space, more metal, more everything. Is then the M5 too lively of ride, too energetic of engine, too, well, *sporting*, when stood alongside a Jaguar V12 saloon? Nonsense, the Jaguar is a cruiser, silky smooth, and with a magic carpet ride, you can't expect it to take bites out of the tarmac whenever you feel moved.

Very well, there's another problem, rather more personal this time. I've met so many people who shake their heads in wonderment whenever the topic of M5 arises. "The most complete car that has yet been invented," wrote one. For them, a few billiard table-smooth miles in the euphoria of a sun-soaked European press launch

Grip, even in the wet, is enormous. Roll is minimal, but lack of steering feedback detracts from the chassis sharpness. M5's power increased by almost 30percent over predecessor (far right).

may have been enough to convince that this BMW can have no peers. Rather like someone telling you that such and such a film just *has* to be seen. Without question it's a masterpiece. For some reason, that kind of assertion tends to make you look more carefully at the subject in hand.

But perhaps it's more relevant to ask whether the attributes of a sports car really can be combined with those of a capacious saloon. A degree of compromise is surely part of the charm of a real sports car. On the other hand, no manufacturer has really yet managed to defy the laws of physics and prop up a big body sufficiently well to be sporting without rattling its occupants, nor overcome the basic problem of making a larger mass change direction with the same agility as a smaller one.

For this particular pressman, the last M5 was the business. It was everything I wanted in a saloon car. Understated, yet sharp and potent, it was rewarding to drive, yet demanded respect without bullying. It had that helping of agility that was still to be fully exorcised from the BMW trailing arm rear suspension; and it had five seats as well.

In this latest aerodynamic version of the M-badged 5-series (a Cd of 0.32 apparently) the number of seats has in fact diminished by one. There is now a luggage compartment which divides the rear bench. Engine power has however gone up. The glorious four-valve-per-cylinder six, which started life in the M1 sports car (and, with two cylinders lopped, forms the M3's

"IS THE M5 TOO LIVELY OF RIDE, TOO ENERGETIC OF ENGINE, TOO, WELL, SPORTING?"

power unit) has retained its 93.4mm bore, but now has an 86mm stroke – an increase of 2mm – which stretches the engine's capacity slightly from 3,453 to 3,535cc. The engine's heavily-oversquare dimensions allow a maximum rotational speed of 7,200rpm – remarkable for a production straight six, a configuration not ideal for high rpm. The 'standard' single-cam six which appears in the 535 has, by comparison, a stroke of 86mm, which with a bore of 92mm gives a swept volume of 3,430cc.

The M5's engine is the biggest capacity six ever produced by BMW, and it now develops a mighty 315bhp, an increase of 29bhp over the last M5, which all now happens at a fairly dizzy 6,900rpm. Torque has risen as well, peaking at 4,750rpm and 266lb ft, and in addition this M5 is the first BMW Motorsport car to feature a full catalytic converter.

More than 300bhp is a lot of power, even when it has 33cwt to motivate, and the M5 lapped the Millbrook bowl at an average of 157mph. This could have been more but for the top speed limiter, which now appears on all BMW's bahn-stormers, and it does work, gently frustrating the car's progress without any jerks or lurches. BMW claims a limited maximum of 155mph, so without that electronic interference, or the tyre-scrubbing curve of the bowl, this car could easily be capable of more than 160mph. This is fast indeed, irrespective of the seating arrangements, and falls only a few miles per hour shy of the Porsche 928S4 SE's monster 164mph top whack. The 928 is but a 2+2, but for the purposes of comparison it probably represents the most *usable* 160mph car we've tested, and of the select few to venture past the 150mph barrier, probably felt the best composed over the bumps of Millbrook's high speed bowl. The BMW came close though, and but for an intermittent roar as the air pressure overcame the driver's window seal, you didn't have much sense of two and a half miles a

minute flashing by. The radius of the bowl and the bumps, coupled with vicious sidewinds at certain points, make for very sticky palms in some cars.

Sticking with the Porsche comparison for the moment, the 928's V8 develops a similar amount of power (320bhp at only 6,000rpm), but boasts a musclebound 317lb ft torque peak at just 3,000rpm, a legacy of 4,900-odd cc no doubt. This shows up in the in-gear acceleration figures, despite a similar all-up weight (33cwt for the BMW, 31cwt for the Pork). The M5's 60-80 and 70-90mph increments take up 5.6sec, the 928 needs 4.2 and 4.6sec respectively.

If the BMW is a sports car, lazy lugging is not what it's about, and if you use the enormous rev range to the full, the M5 will reach 60mph from rest in a mere 5.6sec, and 100mph in 14.6sec which, given the car's size, is remarkable indeed. The engine needs to be worked however, and despite BMW's claim that 80percent of peak torque is available all the way from 3,000rpm, nothing much happens until the rev counter is showing nearly 4,000.

No series production car of similar size can match the BMW for standing start acceleration, or even come close with such a minimum of fuss. The M series engine's extraordinary working range is a matter of wonderment at all times, not least because the usual grit-your-teeth area between six and 7,000rpm can be transcended with no sensation of fear for the reciprocating bits beneath the bonnet. It's not as silky as the 2.5-litre sixes, or quite as smooth as the single-cam 3.5, but that's no bad thing. The hollow growl that chimes in with the onset of real pulling power is heard rather than felt, and it's a pleasant reminder that there's grunt about.

The extended working range of the engine is BMW's pre-emptive answer to any criticisms about lack of power down below, but the practical consequence is that you do always need to be ready to drop two gears in order to blast past a queue. The alternative is to sit in third with the tacho hovering around the 3,500 mark. Response is then devastating, and available to beyond 100mph. Fuel consumption works out at around 17-18mpg.

The gearchange is light, if slightly rubbery while the clutch is heavy, and this particular combination sometimes makes smooth driving a matter of concentration. Ratios are well chosen, and a further consequence of the engine's rev band is that you can hang on to the gears for what seems an inordinate length of time. Second is quite good enough to get you nicked on this blinkered isle.

And so to the thorny question of handling, and the subjective matter of steering feedback. I read somewhere recently a lofty piece by a well-known motoring journalist to the effect that steering feel is irrelevant, and much overrated by the specialist media. The BMW should appeal to this man; either side of the straight ahead the

"IF THE BMW IS A SPORTS CAR, LAZY LUGGING IS NOT WHAT IT'S ABOUT"

steering is rather dead, and merely easing out of the traffic stream needs a distinct turn of the wheel rather than a squeeze of its rim. Once on lock and loaded up, matters improve although there's little sensation of impending loss of grip at the front wheels, of which more later. Maybe the BMW's steering represents a sensible compromise; motorway progress needs no effort to keep the car straight, but when you really start to move, the steering is accurate. It all depends whether this is a sports car or not – which is where we came in. I would rather have a little more instant response when you need it – at the initiation of the manoeuvre. Certainly, for a car of such enormous performance, the M5's is the safe option but the Germans could perhaps learn a little from the French, and Peugeot in particular. Thankfully BMW has resisted the temptation to use the 750's ghastly Servotronic speed-sensitive assistance.

The steering nevertheless improves dramatically when you start to press on. Get the whole process flowing, commit the car to a corner, and the helm immediately feels sharper, the chassis turns as a whole instead of pushing its nose wide, the heavy clutch ceases to matter because you're only kicking at it, the engine sings and you're flying. The car seems to shrink round you and become more agile, and the grip is awesome – on dry roads at least. Beefy 255/45 Michelin MXXs on 9J×17in rims see to that, although tyres of 10in tread width inevitably produce some tramlining over ridges in the road, especially under firm braking. Under pressure, the car makes sense and it's all too easy to find yourself gaining on the traffic in front at an indecent rate. Chassis balance is excellent, a little initial understeer converts to eventual oversteer, which is as it should be, and if you do overdo it on a greasy road it's deliciously simple to hold the slide, retrieve the tail and continue without drama. You do need to take more care when the roads are slippery, not simply because of the amount of power at your disposal – that much is always up to the driver. Rather, the problem arises when

There's little to give away the fact that this is one the world's most potent saloons. Moorland miles were simply gobbled up in miserable conditions.

SPECIFICATION

PRICE	£43,465

ENGINE

CYLINDERS	six, in-line
CAPACITY, cc	3,535
BORE/STROKE, mm	93.4/86
CAMSHAFT	dohc, four valves per cylinder
COMPRESSION RATIO	10.0:1
FUEL SYSTEM	Bosch Motronic
FUEL	unleaded
MAXIMUM POWER, bhp/rpm	315/6,900
MAX TORQUE, lb ft/rpm	265/4,750

TRANSMISSION

TYPE	five-speed manual

INTERNAL RATIOS AND MPH/1,000rpm

FIFTH	0.81:1/23.1
FOURTH	1.00:1/18,8
THIRD	1.35:1/13.9
SECOND	2.08:1/9.0
FIRST	3.51:1/5.3
FINAL DRIVE	3.91:1

CHASSIS

SUSPENSION

Front	double-pivot spring struts with caster angle offset, positive steering scrub radius, lateral force compensation, brake dive reduction
Rear	independent via semi-trailing arms, auxiliary trailing arms, squat and brake dive reduction
STEERING	rack and pinion, power assisted

BRAKES

front/rear	ventilated discs/discs, ABS
WHEELS	alloy, 8J×17
TYRES	235/45 ZR17

DIMENSIONS

LENGTH, in	WIDTH, in	HEIGHT, in	WHEELBASE, in
185.5	68.9	54.8	108.7

TRACK front/rear, in	FUEL TANK, gall	KERB WT, cwt
57.9/58.8	19.8	32.9

PERFORMANCE

MAXIMUM SPEED, mph			157

ACCELERATION THROUGH GEARS, sec

0-30mph	0-40mph	0-50mph	0-60mph
2.3	3.3	4.5	6.5
0-70mph	0-80mph	0-90mph	0-100mph
7.7	9.6	11.6	14.6
0-110mph	0-120mph	0-130mph	
17.7	21.2	25.5	

ACCELERATION IN FOURTH, sec

30-50mph	40-60mph	50-70mph	60-80mph
6.5	6.2	5.8	5.6
70-90mph	80-100mph	90-110mph	100-120mph
5.6	5.7	6.0	6.7

ACCELERATION IN FIFTH, sec

40-60mph	50-70mph	60-80mph	70-90mph
8.4	8.7	8.5	8.2
80-100mph	90-110mph	100-120mph	
8.0	8.2	8.9	

FUEL CONSUMPTION

OVERALL TEST FIGURES, mpg	23.7
GOVERNMENT TEST FIGURES, mpg	
URBAN CYCLE	15.6
STEADY 56mph	34.4
STEADY 75mph	30.1

MAKER/IMPORTER: BMW GB, Ellesfield Avenue, Bracknell, Berkshire, RG12 4TA. Tel: 0344 426565.

you're not trying. On more than one occasion while threading along wet country lanes, the car would understeer quite dramatically without any real provocation. Perhaps we get too used to front-drive scrabblers which will obediently tighten their line when the accelerator is released.

Modifications to the basic 5-series suspension are few; the car sits a little under an inch lower all round, and the anti-roll bars are thicker, by 2mm at the front, 3mm at the rear. The M5's ride is restless, which you might expect, but it's not because the body jiggles in response to bumps, rather that it follows the contours of the road, almost as if the suspension were not moving – which of course it is. Much of this is due to the increased roll stiffness from the thicker bars, and the downside is that it probably doesn't help the wet road understeer. At higher speeds where it really matters, simply superb body control really shows that the engineers have done their homework on the damper settings. Even really clumsy straightening up out of a fast ess-bend won't set the body lurching about.

It's the sort of riding sensation that is usually exclusive to mid-engined cars where the masses are concentrated closer to the centre, and it lends a really classy feel to a big fast saloon. Nothing is more downmarket than a big car which crashes on to its bumpstops whenever you hurry along with a full complement of passengers, and their luggage.

Such immense performance would be irresponsible if the brakes did not match, and at 12.4in × 1.1in thick the new M5's ventilated front discs are bigger than ever, with new floating calipers regulated by ABS. They are indeed powerful, and didn't fade when hauling the car up at the end of the test track's mile straight – the stearnest test outside a racetrack. The feel of the middle pedal, though, is awful. There's too much servo assistance, and the pedal appears to hang on, such that you can't feather off tidily to bring the car to a halt without a rearward lurch. Maybe it's the right-hand-drive conversion, but it's another feature which interferes with the driving enjoyment of a massively capable car.

This really is the crux of the matter. There's no denying the M5's performance, or its handling.

"THE DENIAL OF DRIVING INVOLVEMENT IS A MOAN THAT I HAVE EXPRESSED BEFORE"

Both set new benchmarks for the class. The driver, however, somehow seems yet more isolated from the driving process. The steering is accurate but uninformative; surely this is an instrument which should feed the information about tyre grip to the driver before anything else? The brakes are powerful and fade free, but overservoed and insensitive. The gearchange is fast, but could do with more weight, and the clutch could do with less. The instruments are superb as always, as is the location and operation of the minor controls; the pedals however are well offset to the right.

The denial of driving involvement is a moan that I have expressed before, several times, about the likes of the Porsche Carrera 4, Sapphire Cosworth 4×4 and Lotus Esprit SE. And now the M5. All are more capable than their predecessors, all keep from the driver some of the pleasure in simply operating that the predecessor seemed to have on offer.

The M5 is a very expensive, impressive car that is well made, well thought out and tastefully understated. It's faster and safer than its predecessor. There's nothing so big, and so quick for the money. I just wish I wanted to drive it somewhere simply for the hell of it.

'Some day all cars

Wishful thinking by Dave Calderwood after driving and track testing the latest BMW M3, the 2.5-litre Sports Evolution

Day one

The first time I approach the M3 is in darkness and I can't figure out why the key won't unlock it. It's late and it's been a hectic day at the office so, dozily, I think there must be a security system somewhere for this bright red, limited edition BMW. After looking round the door for an extra lock, I open the boot, half expecting a piercing alarm. Nothing in there, but then I remember the car is left-hand drive — all M3s are. Fool. No alarm, but then a thief would have to get past BMW's deadlocks.

It takes a while to orientate yourself in a left-hand-drive car, especially in right-hand-drive Britain. The gearchange is different, too — a racing pattern with first on a dogleg opposite reverse, with second/third and fourth/fifth opposite one another. The shift is firm, very firm. It needs a solid push to take it out of gear and into the next one. Not slow but firm. It's also difficult to find third and second gears in a hurry when changing down, but that might just be my

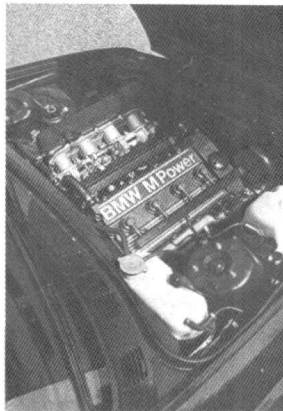

Extra engine capacity gives
M3 more mid-range torque

The front and rear spoilers
each have three settings

will be like this...'

right hand getting used to the left-hand-drive layout.

On the way home, joining the M25 is a nightmare. Vision across the car is patchy; glimpses of a solid stream of traffic filter past the seat and headrest, B pillar and C pillar, confusing the view from the slip road. I'm reminded of a story told by Roger Maingot, managing director of Maranello, the Ferrari concessionaire, about driving the F40. The Ferrari's rear and side vision is so poor as to be non-existent, so he accelerates like hell when using a slip road and hopes he's not going to meet one of his customers doing the same.

Pulling off the motorway as soon as possible on to A-roads, the next impression is just how taut the car feels, and how much more positive the steering is over a standard 3-series BMW. The suspension is firm but not unyielding and there's virtually no body roll. The car never seems to lose its poise and tracks around corners as though on the proverbial rails. The grip from the Michelin MXXs is amazing. I'm going to enjoy this car.

Day two

The temperature gauge takes ages to rise to the middle. I'm watching the needle because I'm stuck in a slow-moving queue on a twisty A-road and I don't want to use all the performance for overtaking until the engine is fully warmed up. Ten miles pass before it's there. Leaving it in third — with plenty of revs left on the tacho — I edge out for a look. This takes a lot of care and cunning positioning since left-hand drive hinders again.

On a long, sweeping right-hander, I can see far enough ahead to come out for a good look. It's clear. Gas it and at first, with the revs down around 2500, it's slow to pick up. Then, as 3000rpm comes up, the gentle but steady acceleration turns into a rush. It's not so much a kick in the pants as a firm push back into the seat. The tacho needle shoots round to the last 1000rpm before the redline at 7000rpm and suddenly the car is very busy indeed.

We've overtaken the queue, nipped back into lane and are

BMW M3 SPORTS EVOLUTION

All tests with a crew of two and a full tank of fuel

THROUGH THE GEARS (seconds)

0-30mph	2.3	**0-70mph**	8.0
0-40mph	3.3	**0-80mph**	10.1
0-50mph	4.8	**0-90mph**	12.1
0-60mph	6.1	**0-100mph**	16.0

STANDING 1/4 MILE	14.7sec
TERMINAL SPEED	95.5mph
AVERAGED TOP SPEED	149mph
FASTEST 1/4 MILE	154mph

ACCELERATION IN 4th/5th (seconds)

30-50mph	6.4/9.3	**60-80mph**	6.8/10.3
40-60mph	6.6/8.4	**70-90mph**	7.1/10.9
50-70mph	6.4/8.8	**80-100mph**	6.9/11.3

MAX SPEED IN GEARS @ 7000rpm

FIRST	42mph	**FOURTH**	123mph
SECOND	65mph	**FIFTH**	149mph @ 6715rpm
THIRD	88mph		

OVERALL FUEL CONSUMPTION	25.7mpg/5.6mpl
PROVING GROUND CONSUMPTION	11.4mpg/2.5mpl
GOVT FIGURES	Urban 22.8mpg; @ 56mph 45.6mpg; @ 75mph 36.2mpg
TRACK CONDITION	Slightly damp, drying
TEMPERATURE	+14°C
WINDSPEED	8-12mph
ATMOSPHERIC PRESSURE	758mb

SPECIFICATION

ENGINE TYPE	In-line, longitudinal
DISPLACEMENT	2467cc
BORE	95mm
STROKE	87mm
COMPRESSION RATIO	10.2:1
MAX QUOTED POWER (DIN)	238bhp @ 7000rpm
MAX QUOTED TORQUE (DIN)	177lb ft @ 4750rpm
BHP PER LITRE	96.5
POWER TO WEIGHT RATIO (UNLADEN WEIGHT)	201.5bhp/ton
POWER TO WEIGHT RATIO (TEST WEIGHT)	170.3bhp/ton
FUEL SYSTEM	Digital Motor Electronics injection
CYLINDER HEAD	Alloy, twin overhead camshafts, four valves per cylinder
GEARBOX	Five-speed manual

GEAR RATIOS

TOP	1.00	**2nd**	2.40
4th	1.26	**1st**	3.72
3rd	1.77	**Reverse**	4.23

FINAL DRIVE RATIO	3.15
FRONT SUSPENSION	Independent by MacPherson struts, coil springs, lower wishbones and anti-roll bar
REAR SUSPENSION	Independent by semi-trailing arms, coil springs and anti-roll bar
BRAKES	Single piston floating caliper, vented discs, ABS
STEERING	Speed-sensitive power assisted, rack and pinion
WHEELS, TYRES	7½J x 16in, Michelin MXX 225/45
UNLADEN WEIGHT	2646lb
TEST WEIGHT WITH CREW AND EQUIPMENT	3131lb
WHEELBASE	101.0in
TURNING CIRCLE	36.4ft
FUEL TANK CAPACITY	13.6 gallons/62 litres
BASIC PRICE (INC TAXES)	£34,500
PRICE AS TESTED	£37,155

OPTIONAL EXTRAS FITTED TO TEST CAR leather seats £1310, sun roof £745, electric front windows £413, radio fitting service £187

'Some day all cars will be like this. . .'

proceeding along at a very healthy rate, with the engine note somewhere between a throaty growl and a sewing machine on full chat. The engine feels smooth even at these revs and I can't help making the comparison with Ford's Cosworth lump — both highly-tuned four-cylinder 16-valve units, but the rough Cossie always gives the impression of being stressed.

Today we're heading for Millbrook, the motor industry proving ground where we can test the car for ultimate performance figures. This Sports Evolution version of the M3 has been bored and stroked to give an extra 165cc, and has been fitted with slightly bigger inlet valves (wider by 0.5mm) and an intake camshaft with a longer opening time. Key hot spots receive help: the exhaust valves are sodium-filled and the pistons are cooled by new oil jets. The changes benefit both power (up 11 per cent to 238bhp at 7000rpm) and torque (up four per cent to 177lb ft at 4750rpm). All this with a metal catalytic converter and on 95-octane unleaded fuel. We fill it up with Supergreen, even though BMW says it makes no difference.

First, top speed. Around the top lane of the two-mile banked bowl in fifth gear, the revs edge higher and higher, eventually settling between 6600 and 7000rpm, depending on whether we're heading into the wind. A lap time of 48.99 seconds equates to a speed of 149mph, while the fastest quarter mile, when we're in the lee of the hill, equates to 154mph — exactly what BMW claims.

It's hard work on the bowl at this speed, because there's a lot of tyre scrub — indicated by the right-arm pressure required to steer down the banking in order to stay in a straight line — and the wind buffets us around a bit. One of the door seals isn't coping with the speed and it sounds as though a gale is blowing.

On the mile-long straights where we test for acceleration, the Sports Evolution amazes us. Leaving big black tyre footprints, it scorches to 60mph from rest in 6.1 seconds, to 100mph in 16 seconds flat and dispatches the quarter mile in 14.7 seconds. It's the fastest M3 yet, and judging by the times, I've got used to the awkward gearchange pattern.

The session at Millbrook overcomes my nervousness with the car and the drive home is fast, the M3 feeling completely competent. The grip is always there, no matter how much you demand of the car. It's such a 'chuckable' car, too; the steering is neither light nor heavy. Self-centering is fairly strong, offering enough resistance to make positioning accurate, yet the wheel encourages you through twisty switchbacks where you're constantly changing direction. However, it's a bit slow to turn in (with 3.7 turns from lock to lock), which is the only disappointment.

That evening, we meet friends for dinner in Chiswick. The M3 is conspicuous even in yuppie West London. Spoilers front and rear shout 'racer' to the world and bulging wheel arches add to the effect. Each spoiler blade has three settings — to reduce lift and add downforce — giving nine possible settings for racing. For Chiswick High Road, I opt for the middle setting on both. It's the right one, I'm sure.

Day three

Office day today, but others take the M3 out for a spin and come back enthusing. It's the grip, again. There's just no way those tyres are going to come unstuck in road use, in the dry. John Barker notes that the torquier engine means you spend less time trying to keep the engine on the boil and more time enjoying the handling. Worryingly, he also says that the engine rev limiter is so abrupt that the first time you hit it, it feels as though the engine has blown up.

The homeward journey is slow, because a smattering of rain brings traffic to a standstill at strategic points. Never mind; it's an opportunity to study the interior of the M3. Almost every square inch is covered in taut matt black leather with big bold Motorsport stripes across the front seats. These are real racing items that hold you in place no matter what the cornering forces are. Bright red diagonal seat belts complete the security, and a brushed suede steering wheel rim — in black, of course — has an almost sensuous touch. The Sports Evolution is fitted with all mod cons and is surprisingly easy to drive in stop-start traffic.

Day six

Where did days four and five go? Others in the office demanded to

drive the M3, and in the interests of office harmony, the keys are handed over in exchange for a Toyota Corolla. That was the last I saw of it until deputy editor Howard Lees came back enthusing about the grip, again, and pointing out that the slight understeer gives a healthy safety margin. Good, I say, I'm going to need it for my final day with the car.

Given the Sports Evolution's *raison d'etre* , a session at a racing circuit is deemed necessary. After an exhilarating early-morning drive across country, I arrive at Castle Combe, adrenalin all pumped up for a track I've never seen before. For a circuit that's been in use for 40 years, it's a low-key place and very straightforward, apart from one bend, Quarry. The corner itself is all right — a near-hairpin taken in third — but the approach is far from straightforward.

Along the start/finish straight, you're accelerating hard in third, just snicking into fourth before Folly, a right-hand kink taken with the right foot hard down. Pull the car back over to the right side of the wide track, slightly uphill and undulating here, before braking hard in the slight dip of Avon Rise. This is a left-hander immediately before the right-handed Quarry, so you enter late to avoid drifting to the right which would put you off line for Quarry.

The M3 is brilliant here, allowing me to cock it up several times and experiment with lines over the bumps and undulations. Whoever designed this corner had one thing in mind — to upset the car's stability before Quarry. The left-hander of Avon Rise is just where you should be braking hard for Quarry but you can't — you'd lose the back end over the hump of the turn. And in between this

> Even when I get it all wrong and arrive at Quarry with tyres screaming, in the wrong gear, the M3 soaks up the punishment and allows me to muddle through

and the apex of Quarry you have to change down too. Even when I get it all wrong and arrive in the middle of Quarry with tyres screaming, in the wrong gear and understeering like mad towards the edge of the track, the Sports Evolution soaks up the punishment and gives me time to muddle through. No matter how hard the brakes are used, they always work, and a new pad material resists heat fade admirably.

Eventually I work out a line that maximises everything — to my level of skill, that is — and the M3 allows me to chip away at the lap time, getting smoother and faster in an even, progressive fashion. It's one of the few cars that makes both a great road car and a great racer. One of a few is right — just 50 Sports Evolutions will come to Britain from a production total of 600. They're being built to homologate the new engine and body parts, as well as a revision to the suspension that lowers the front by 10mm. The flared wheel arches allow the use of wider tyres for racing, according to BMW, and the Sports Evolution comes with 225/45 ZR16s, compared with the standard M3's 205/55 ZR15s.

There are other, less noticeable details, too, such as the modified fin profile on the kidney grille for better cooling, and all the gaps at the front are sealed for better air flow. The front and rear bumpers are lighter, as are the boot lid and the rear and side windows. A bigger boot is a benefit of a slightly smaller fuel tank (13.6 gallons), which still gives a range of 350 miles at an average consumption of 25.7mpg. The session at Castle Combe dropped the consumption down to 12mpg, which roughly equates to twice as much fun as on the road. Yes, that's about right.

Sport and touring disciplines are inherently conflicting. Sports cars need to be light, responsive, and firmly suspended; luxury cars soft, quiet, and smooth. Perhaps more than any other car company, BMW has demonstrated the ability to fuse successfully touring and sport technologies to offer vehicles that do both well.

The 7-series Beemers are benchmark tourers, and not without some good handling qualities. Likewise, the 5-series cars offer exemplary handling, combined with more than acceptable traveling credentials. Sitting at the top of the 5-series lineup is the new-for-'91 M5. At the top of the 7s is the much improved 750iL.

The M5 is for the driver who wants a second-to-none performance weapon that can be pressed into service for a long weekend trip to the ski lodge.

The 750iL driver focuses on luxurious travel with all the amenities, but when the other seats are empty, you'll find him on those interesting back roads, with his foot well into the wonderful V-12. Since the M5 and 750iL bring such different approaches to motoring from the same company, we thought they'd make an interesting case study.

BMW's first M (motorsport) product was the M1, a two-seat sport GT with the same basic 3.5-liter 24-valve inline six found in the new M5. As you would expect, the modern M5 version is much more sophisticated. An earlier version of the M5 was offered in 1987 and 1988, but was dropped when the new 5-series was introduced in 1989.

Back for '91, the M5 offers significant changes in the engine compartment, the result of which is an outstanding 310 horsepower. The '91 engine has a 2-millimeter stroke increase, which ups displacement to 3535 cubic centimeters. Piston changes have bumped compression to 10.0:1, changing the fuel requirement to premium unleaded. The intake system now uses an electronically controlled dual resonance system for good midrange torque, hot-wire airflow sensing for better response, refined intake ports and plumbing, and new camshaft specs with increased valve lift.

On the exhaust side of the engine, engineers fitted an equal-length stainless steel header system, which feeds into a large-volume catalytic converter and muffler taken directly from the 750iL parts bin. A new forged crankshaft raises safe rpm to 7200 to accommodate the higher-lifting, higher-revving cam specs.

The engineer's work has paid off; 80

percent of maximum torque is available from 2900 rpm all the way to redline. In other words, 80 percent or more of max torque is available over 75 percent of the useful rev range of the engine. This engine *defines* flexibility. In addition, EPA mileage is 11/20, city/highway, acceptable for such a high specific output engine, and up from the earlier M5, in spite of the increased power.

To make sure all that lovely torque is transmitted to the road properly, the chassis engineers tweaked the chassis accordingly. The M5 is 20 millimeters lower, has increased anti-roll bar diameters at both ends (to 25 millimeters in front, to 18 millimeters at the rear), and firmer calibration for all four gas shocks. A new self-leveling system at the rear alleviates fears of overheating the inside of the tires in a heavily loaded car at high speeds. Steering ratio is reduced 15 percent and effort increased slightly.

A car of this potential needs good brakes, and the Motorsport Group responded willingly. The big 12.4-inch ventilated discs are fed by the special two-piece wheel design, which flows air axially and radially through the brake disc for maximum cooling. A 25-percent limited-slip differential is standard on the M5. The final factor in a car that handles well is tires, and the M5 is blessed with a set of Michelin MXX2s, sized at a whopping 235/45ZR17—this is serious rubber.

The M5 has been fitted with side skirts and front and rear spoilers. These combine to give the M5 the same drag coefficient as the 535, in spite of the much larger tires and increased cooling air demand. The body aerodynamics also reduce lift at both

BMW M5 VS. BMW 750iL

by Ron Grable
PHOTOGRAPHY BY SCOTT KILLEEN

THE SPORT/LUXURY APPROACH TO MOTORING

axles, compared to the 535. The only other significant change for '91 is that BMW fitted a bench seat in the rear to make the M5 a true five-seater.

The heart and soul of the 750iL is unquestionably the velvet-smooth 5-liter V-12. For '91, the four-speed automatic has been revised to provide earlier downshifts at partial throttle to enhance acceleration and response. The engine control computer, called DME (digital motor electron-

ics), has been programmed with new maps for control of fuel and spark. This provides increased fuel efficiency, even with the new-for-'91 lower final-drive ratios.

Upgrades to the anti-lock brake system reduce yaw inputs when stopping on surfaces with varying friction and provide a smoother pedal response when the ABS is cycling. ASC (automatic stability control) is now ASC+T (traction). The earlier ASC system controlled throttle, spark, and fuel to limit wheelspin on slippery surfaces. The +T incorporates rear wheel braking to provide quicker response to incipient wheelspin.

An additional subtle feature of +T is that it stabilizes the vehicle during deceleration on slippery surfaces. Under such conditions, quickly closing the throttle or an automatic downshift can lock the rear wheels from engine braking, and the ASC+T function acts to restore rolling motion to the wheel with throttle, fuel injection, or spark. Starting in 1991, all 7-series BMWs have a telescopically adjustable steering wheel and auto-dimming rearview mirror.

A new BMW option for '91 avail-

TECH DATA

GENERAL		
	BMW M5	**BMW 750iL**
Distributor	BMW of North America, Inc., Woodcliff Lake, N.J.	BMW of North America, Inc., Woodcliff Lake, N.J.
Body style	4-door, 4-passenger	4-door, 5-passenger
Drivetrain layout	Front engine/rear drive	Front engine/rear drive
Base price	$56,600	$74,000
Price as tested	$59,055	$76,105
Options included	Dealer prep, $230; destination, $375	Dealer prep, $230; destination, $375
DIMENSIONS		
Wheelbase, in./mm	108.7/2760	116.0/2945
Track, in./mm	58.0/58.9/1475/1495	60.2/61.3/1530/1557
Length, in./mm	185.8/4719	197.8/5024
Width, in./mm	68.9/1750	72.6/1844
Height, in./mm	55.4/1407	55.1/1400
Ground clearance, in./mm	4.6/116	4.7/120
Mfr's curb weight, lb	3804	4235
Weight distribution, f/r	50/50	51/49
Cargo capacity, cu ft	16.2	17.6
Fuel capacity, gal	23.8	24.0
Weight/power ratio, lb/hp	12.3	14.3
ENGINE		
Type	Inline 6, liquid cooled, cast iron block, cast aluminum head	V-12, liquid cooled, cast aluminum block and heads
Bore x stroke, in./mm	3.68 x 3.39/ 93.4 x 86.0	3.31 x 2.95/ 84.0 x 75.0
Displacement, ci/cc	216/3535	304/4988
Compression ratio	10.0:1	8.8:1
Valve gear	DOHC, 4 valves per cylinder	SOHC, 2 valves per cylinder
Fuel/induction system	EFI	EFI
Horsepower, hp @ rpm, SAE net	310 @ 6900	296 @ 5200
Torque, lb-ft @ rpm, SAE net	265 @ 4750	332 @ 4100
Horsepower/liter	87.7	59.3
Redline, rpm	7000	6000
Recommended fuel	Unleaded premium	Unleaded premium

able on 750iL, 735iL, 735i, and 850i is Electronic Damping Control. EDC has three levels of shock valving: soft, medium, and firm. A computer chooses valving based on longitudinal acceleration, vertical wheel motion, steering wheel angle, and vehicle speed. This is much more sophisticated than the average "automatic shock damping" system. It uses magnetic switching to get instantaneous changes between damping levels, and a 16-bit 12-megahertz microprocessor to execute the damping rate changes in less than a half second.

This system doesn't firm up the shocks as a function of vehicle speed; it changes when the road or dynamic conditions demand it as determined by the sensors. This system has allowed engineers to choose a damping rate for the soft mode that's only 30 percent of the normal 750iL shocks. We're talking plush here. The driver can also manually select Comfort or Sport from a console switch. The Sport setting simply eliminates the soft mode.

Driving these two thoroughbreds is a study in contrasts. For example, handling of both cars is nearly flawless (in context), but each does its job in its own way. The M5 is aggressive and precise, with enough immediate power to allow cornering attitude control with the right foot. On the other hand, the 750iL doesn't feel exactly comfortable generating 0.81 gs, or transitioning from corner to corner. It's almost as if it *can* do it—but wouldn't it be so much better to just lean back and call up some stereophonic rhapsody from the trunk-mounted CD player?

Any cornering short of the limit is easily controllable in the 750iL, and the 5-liter V-12 devours straights between corners in an effortless, almost

CHASSIS		
	BMW M5	**BMW 750iL**
Suspension		
Front	MacPherson struts, lower control arms, coil springs, anti-roll bar	MacPherson struts, lower control arms, coil springs, anti-roll bar
Rear	Semi-trailing arms, coil springs, anti-roll bar, self-leveling	Semi-trailing arms, coil springs, anti-roll bar, self-leveling
Steering		
Type	Recirculating ball, power assist	Recirculating ball, power assist
Ratio	15.6:1	16.2:1
Turns, lock to lock	3.3	3.5
Turning circle, ft	36.1	39.4
Brakes		
Front, type/dia., in.	Vented discs/12.4	Vented discs/12.8
Rear, type/dia., in. Anti-lock	Discs/11.8 Standard	Vented discs/11.8 Standard
Wheels and tires Wheel size, in. Wheel type/material Tire size Tire mfr. and model	17 x 8.0 Forged aluminum 235/45ZR17 Michelin MXX2	15 x 7.0 Cast aluminum 225/60VR15 Pirelli P600
DRIVELINE		
Transmission type	5-spd man.	4-spd auto.
Gear ratios (1st) (2nd) (3rd) (4th)	3.51:1 2.08:1 1.35:1 1.00:1	2.48:1 1.48:1 1.00:1 0.73:1
Axle ratio	3.91:1	3.15:1
Final-drive ratio	3.17:1	2.30:1
Engine rpm, 60 mph in top gear	2600	2050

contemptuous whoosh. The 750iL and 735 share chassis and suspension components, and the 750 handles nearly as well as we've come to expect from the 735. Press out near the cornering limits, and the 750 just progressively understeers more, remaining stable and predictable—exactly the characteristics BMW's engineers designed into it.

The M5 is another story altogether. Each muscle-bound horsepower is responsible for 2 fewer pounds than the 750.

This superior power-to-weight ratio produces a marvelous ferocity that slams you into the seat at the slightest provocation. The engine pulls hard anywhere above 3000 rpm and feels happiest around redline.

The chassis dynamics of the M5 are the stuff of legends. Who cares if it's a little harsh on the freeway? The M5 is to a curvy mountain road what Kenny Roberts is to a motorcycle, Heifetz to a violin, and Yeager to a jet. It's perfectly balanced for fast motoring and remains stable and linear right out to its impressive limits. The brakes are like a carrier arrester wire and practically fade-free. BMW sport seats provide the perfect platform for the job of high-performance driving. They're firm and keep the driver placed behind the wheel with a minimum of effort.

While we had these two in our possession, we couldn't resist a fast run with both cars on the high oval at Calsonic's high-speed test track in Arizona. Even though both have artificially limited top speeds of 155 mph, we wanted to see how stable they are at the upper end. We weren't disappointed; both were at ease and unflappable on the banking. The noise level inside the 750 was phenomenally low at speed, with only a muted pulsation emanating from the front.

The M5 was equally stable, but felt hyper. The mesmerizing engine note was a combination of an intake air moan and the whir of complex mechanicals doing their designated stuff. Wind noise was a little more noticeable and the intense chassis tuning required a higher level of driver concentration.

Each BMW offers an abundance of goodies for the customer. The 750iL essentially has no competition in its class. That'll change when the Mercedes-Benz 600SEL with the V-12 engine appears here, but the Merc will sell for $90,000 plus, quite a bit more than the 750iL. The M5 is equally difficult to target, the closest competitor being the Audi V8 five-speed. It's 0.3 seconds slower 0-60 mph, almost equal on the skidpad, and has lower pricetag, but it doesn't have the same character as the M5. The 500E Mercedes will be every bit the M5's equal when it gets stateside in 1991.

So for now, both BMWs enjoy a unique position, with no direct competition. Comparing the two to each other, the M5 is faster, the 750iL smoother at roughly $20,000 more. We like the M5 because of its character and racing heritage. It's so fast it makes your eyes water, and that's enough for us. **MT**

PERFORMANCE AND TEST DATA

	BMW M5	BMW 750iL
Acceleration, sec		
0-30 mph	2.3	3.3
0-40 mph	3.8	4.8
0-50 mph	5.1	6.3
0-60 mph	6.5	8.1
0-70 mph	8.9	10.4
0-80 mph	11.1	12.8
Standing quarter mile, sec @ mph	14.9/97.5	16.0/91.8
Braking, ft		
30-0 mph	30	33
60-0 mph	120	122
Handling		
Lateral acceleration, g	0.81	0.73
Speed through 600-ft slalom, mph	66.5	57.7
Speedometer error		
Indicated/actual	30/28	30/29
	40/38	40/39
	50/48	50/49
	60/58	60/59
Interior noise, dBA		
Idling in neutral	57	48
Steady 60 mph, top gear	63	63

INSTRUMENTATION

	BMW M5	BMW 750iL
Instruments	180-mph speedo; 8000-rpm tach; fuel level; oil pressure; coolant temp; clock	170-mph speedo; 8000-rpm tach; fuel level; oil pressure; coolant temp; clock
Warning lamps	Oil; battery; ABS, brake fluid; airbag; check engine;	Oil; battery; ABS; brake fluid; airbag; check engine

FUEL ECONOMY

	BMW M5	BMW 750iL
EPA, city/hwy., mpg	11/20	12/18
Est. range, city/hwy., miles	262/476	288/432

BMW M5

An exceptional car for a select few

SOME 25 YEARS ago, Bayer-ische Motoren Werke pop-ularized the idea of the medium-size, high-perfor-mance sports sedan. And there's a direct lineage from the 1800 TIs of those days through the 2500s, the Bavarias and the first-generation 5-Series BMWs to the car you see here, the M5.

Nor is its M heritage any less tell-ing. In 1972, BMW Motorsport GmbH was formed, giving focus to the company's many competition in-terests. World Touring Car Champi-onships and plenty of other honors followed, not to say some pretty po-tent automobiles. In 1978, for in-stance, Motorsport brought forth the M1, a mid-engine Group 4/Group 5 racer whose obligatory 400-car pro-duction run for reasons of homolo-gation gave us one of the all-time great exotics. (See September 1980 for our test of the M1—and remem-ber its dohc 24-valve engine.) Others earning the red, blue and purple BMW Motorsport insignia include the M635CSi (September 1984), the M535i (May 1985), the first-genera-tion M5 (April 1986) and the M3 (February 1988).

BMW Motorsport keeps busy.

So when the latest 5-Series BMW showed up a couple of years ago, we hoped its M variant wasn't far off. And here's that very car, available to a select few North Americans as a 1991 model.

At first glance, only the cogno-scenti will distinguish an M5 from its 535i sibling; and the M5 driver prob-ably wouldn't want it any other way. A revised air dam up front, different bumper contours, subtle sidesills and 0.8-in. less ride height all con-tribute to retaining the 535i's 0.32 C_X, despite the M5's greater intake of cooling air and the increased fron-tal projection of its wide, super-low-profile 235/45ZR-17s mounted on 8-in. wheels.

These wheels, by the way, generat-ed our only M5 style controversy.

Actually 5-spoke skeletal structures with bolt-on magnesium inserts, the latter have concentric rotor blades that cool the M5's oversize ABS-augmented brakes. What's more, the wheels' asymmetric rim design improves run-flat capability.

Elegant engineering, yes; but, as one driver said, "The wheels look too delicate and effeminate for the car's image. Plus, their rims and openings combine to give the appearance of tall-profile whitewalls."

No one objected to other aspects of the M5's suspension, though. Compared with the 535i's, its springs and shocks are stiffened. A limited-slip differential is fitted. Front and rear anti-roll bars are enlarged, the latter just a tad more than proportionally to bring the M5's fore/aft balance a bit closer to neutral.

"Taut but supple," noted one driver, "and at least one Japanese manufacturer could take lessons here in jiggle-avoidance."

"What wonderful pointability," said another, in his admiration of the M5's willingness to be placed precisely with throttle and steering, seemingly regardless of road surface.

Exemplary though the handling is, what characterizes the M5 best is its marvelous powerplant. This inline-6 has double overhead camshafts, 24 valves, a 10.0:1 compression ratio, Bosch Digital Motor Electronics engine management— and a mere, slight, tiny edge of mechanical harshness. Remember that BMW M1 exotic? This engine is a direct descendant of its powerplant. And as it burbles with just a bit of lumpiness at idle or wails like a banshee to its 7200-rpm redline, there's no mistaking its sporting heritage.

Refinements along the way include the Bosch DME combining ignition and fuel control, whereas the original M1 engine had mechanical injection. Also, new to our M5 are a forged crankshaft giving increased stroke and 3535-cc displacement, hitherto 3453 cc, as well as a lightened flywheel, revised camshaft profiles and resonance-charged intake passages and exhaust ducting.

Its resulting 310 bhp and 265 lb.-ft. of torque are produced with no loss of tractability in the sort of driving most of us do most of the time. Yet the M5 is capable of reaching 60 mph in 6.4 seconds, the quarter-mile marker in 15.0 sec. at 96.0 mph and, ultimately, a top speed (155 mph) limited solely by a consensus of German government and industry. Not bad for a luxurious 4-door sedan.

And luxurious the M5 is, in a Bavarian manner that's not as austere as Mercedes-Benz' philosophy, but still unmistakably Germanic. The interior is composed of subtly interacting surfaces, none of the organic wholeness typifying the latest of Japanese design. The driver faces an array of analog dials, black on white, and an airbag-fitted steering wheel. His or her passenger gets a glovebox whose ample volume rivals that of many apartment closets. Between them is a center console canted toward the driver and housing controls for climate (split left/right) and sound system.

Leather upholstery is standard, with exactly four seating positions defined,

■ Without glancing under the hood at Motorsport's twincam inline-6 or spending some time behind the M5's airbag-equipped wheel, most would never know this sleeper packs 310 rear-drive bhp.

the rear pair separated by a fixed center console enclosing a slide-out tray for cassette tapes and the like.

The overall feeling is that of a 4-place cockpit: functional and comfortable, if not particularly spacious.

Apart from its not being the preferred conveyance for four basketball players, why wouldn't just about anyone else absolutely lust for the BMW M5?

"The car's personality comes through loud and clear," said one staff member, "but there are some annoying aspects. Despite all the electric adjustments, I still can't find a decent driving position. The seat has too much lumbar support and the steering wheel is too high, excessively horizontal and not adjustable. Before airbags, BMW had an adjustable wheel, and others with airbags still do."

"The shifter," noted another, "took BMW's characteristically notchy, long-throw feeling to an extreme, especially in 1st–2nd actuation."

The point, of course, is that the M5 isn't a car for everyone. It hasn't been "clinic-ed" to mass acceptability or, worse, to mediocrity. Rather, it's a rolling tribute to BMW confidence, designed by Bavarian engineers who are damned sure of what they know. The M5 works splendidly for some and, we recognize, it just doesn't work at all for others (especially at the price).

Is it worth $56,600? Certainly to a select few. And to the rest, it can still serve as a comforting example of continuity: Considering that the Bavarians all but invented the sports sedan, is it any wonder they continue to do it so well and with such personality? ⊚

BMW M5

0–60 mph	6.4 sec
0–¼ mi	15.0 sec
Top speed	est 155 mph
Skidpad	0.80g
Slalom	61.4 mph
Brake rating	excellent

PRICE

List price, all POE ....**$56,600**   Price as tested **$59,655**
Price as tested includes standard equip. (air cond, AM/FM stereo/cassette, ABS, leather interior, cruise control, electric window lifts, sunroof, central locking & adjustable mirrors), gas-guzzler tax ($1850), cellular telephone ($1205).

ENGINE

Type	4-valve/cyl dohc **inline-6**
Displacement	216 cu in./3535 cc
Bore x stroke	3.68 x 3.39 in./93.4 x 86.0 mm
Compression ratio	10.0:1
Horsepower (SAE):	**310 bhp @ 6900 rpm**
Torque	**265 lb-ft @ 4750 rpm**
Maximum engine speed	7200 rpm
Fuel injection	Bosch Motronic elect. port
Fuel	prem unleaded, 91 pump oct

GENERAL DATA

Curb weight	**3950 lb**
Test weight	**4060 lb**
Weight dist, f/r, %	**50/50**
Wheelbase	108.7 in.
Track, f/r	58.0 in./58.9 in.
Length	185.8 in.
Width	68.9 in.
Height	55.4 in.
Trunk space	17.5 cu ft

DRIVETRAIN

Transmission ....**5-sp manual**

Gear	Ratio	Overall ratio	(Rpm) Mph
1st	3.51:1	13.72:1	37
2nd	2.08:1	8.13:1	62
3rd	1.35:1	5.28:1	96
4th	1.00:1	3.91:1	130
5th	0.81:1	3.17:1	est (7070) 155

Final drive ratio 3.91:1
Engine rpm @ 60 mph in 5th 2740

CHASSIS & BODY

Layout	**front engine/rear drive**
Body/frame	unit steel
Brakes, f/r	**12.4-in. vented discs/11.8-in. discs;** vacuum assist, ABS
Wheels	forged alloy, **17 x 8J**
Tires	Michelin MXX2, **235/45ZR-17**
Steering	**recirculating ball,** power assist
Turns, lock to lock	3.3
Suspension, f/r:	**MacPherson struts,** lower lateral links, compliance struts, coil springs, tube shocks, anti-roll bar/**semi-trailing arms,** coil springs, tube shocks, anti-roll bar

ACCELERATION

Time to speed	Seconds
0–30 mph	2.5
0–60 mph	6.4
0–80 mph	10.7
Time to distance	
0–100 ft	3.2
0–500 ft	8.3
0–1320 ft (¼ mi)	15.0 @ 96.0 mph

BRAKING

Minimum stopping distance	
From 60 mph	127 ft
From 80 mph	231 ft
Control	excellent
Pedal effort for 0.5g stop	14 lb
Fade, effort after six 0.5g stops from 60 mph	19 lb
Brake feel	excellent
Overall brake rating	excellent

FUEL ECONOMY

Normal driving	15.5 mpg
EPA city/highway	11/20 mpg
Fuel capacity	23.8 gal.

HANDLING

Lateral accel (200-ft skidpad)	0.80g
Balance	mild understeer
Speed thru 700-ft slalom	61.4 mph
Balance	mild understeer

INTERIOR NOISE

Idle in neutral	58 dBA
Constant 70 mph	68 dBA

Test Notes . . .

■ The M5 has that rare quality of being not only fast, but also able to accelerate to high speeds with absolutely no drama. In fact, it can be so quiet and smooth that you can hardly believe the speedometer.

■ Our handling measures of grip and control—lateral acceleration and slalom speed—indicate only average road qualities. However, subjectively, the M5 is remarkably well balanced, with agile handling and rewarding steering feel.

Subjective ratings consist of excellent, very good, good, average, poor.

ACCIDENT
OF HISTORY

The BMW M1 Procar series fired passions a decade ago, in the way that the XJR-15 challenge is doing now. Born out of a failed race project, it was a huge success nonetheless. By Alan Henry

M1 Procars engaged in close, sometimes unruly racing. The first five cars on each grid were piloted by top F1 drivers

WHAT DO MCLAREN BOSS RON Dennis and Jaguar's racing overlord Tom Walkinshaw have in common that specifically links them with grand prix racing? The answer to this brain teaser is that, come next week's Monaco Grand Prix, they will have both made their racing debuts in the F1 pit lane fielding sports cars prepared by their own companies.

How come? Well, when Walkinshaw turns up at Monaco to usher out his fleet of delectable Jaguar XJR-15s for the first of the three rounds to qualify for the JaguarSport Intercontinental Challenge, it will be 12 years since Dennis's Project 4 organisation fielded a Marlboro-backed BMW M1 for Niki Lauda to drive in the 1979 Procar series.

If the JaguarSport challenge turns out to be half as riveting as Procar was, then this year's Monaco Grand Prix will be firmly shoved into the shade. One-make racing has frequently provided tedious processions, but Procar — accident of history though it was — provided some of the closest, most hair-raising racing seen in Europe during the summers of 1979 and 1980.

Cynics would say that the advent of Procar saved Jochen Neerpasch's bacon as BMW competitions manager. The mid-engined BMW M1 had originally been manufactured not merely as a luxury supercar to enhance the Munich firm's image, but as the basis of a Group 5 racer with a view to challenging Porsche's supremacy in endurance racing.

Unfortunately, that project fell flat on its face, largely because the sportscar racing rules were changed during the M1's long gestation period. A Group 5 version was eventually produced but it never proved sufficiently powerful match the Porsche 935.

To give the M1 a competition function in life, Neerpasch and Max Mosley (Bernie Ecclestone's right-hand man in the Formula One Constructors' Association) came up with the Procar concept. Some purists were offended by the concept; BMW Motorsport's Denis Jenkinson branded it a "circus act" and Ken Tyrrell expressed the view that if BMW wanted publicity on the back of the F1 circus, it ought to get on and build its own F1 car.

In a whirlwind of publicity, the Procar programme was formally unveiled at the start of the 1979 season. Rounds of the new championship were confirmed as supporting events at the Belgian, Monaco, French, British,

'An enraged Stuck recovered to finish third, after which he had to be restrained as he vented his fury on the winner'

German, Austrian, Dutch and Italian grands prix and a total of 26 cars was expected to battle for the 25 places on the grids.

The first five places on each grid were reserved for the five fastest F1 drivers available (some were excluded through contractual obligations) from the grand prix grid. These machines were built and entered by BMW Motorsport. The others were 'customer' cars, most of which, as things turned out, were manufactured by Ron Dennis's company. Project 4 fielded Niki Lauda's car throughout 1979 and eventually built a total of 30 over two seasons. The customer cars were driven by former or non-F1 personalities such as Hans Stuck, Toine Hezemans, Eddie Cheever, Marc Surer and Manfred Winkelhock. A couple were even prepared by Tom Walkinshaw Racing, originally designated for Dieter Quester and Derek Bell to drive.

With 470bhp available from their 24-valve, six-cylinder 3.5-litre engines, the Procar BMW M1s were serious pieces of machinery and invariably provided fast, close and at times unruly racing.

But the first Procar race was something of a fiasco, with most of the guest drivers retiring and Elio de Angelis emerging the winner from Hezemans and Clay Regazzoni. More interesting than the result was the opportunity to see how the F1 drivers handled this highly competitive sideshow.

Some tackled it with superficial brio, al-most as if they were being offered a specially sanctioned opportunity to let off steam on the eve of a grand prix. Others, notably Lauda, approached it in a professional and meticulous manner. At the end of season, it was no surprise that Lauda and Dennis clinched the championship with wins at Monaco and Silverstone, plus a string of strong placings. Looking back, it almost seems like a dry run for their world championship-winning partnership five years later.

The Procars eventually strayed beyond the confines of a grand prix supporting programme. A non-championship F1 race had been planned at Donington Park in May 1979 in aid of the Gunnar Nilsson cancer treatment campaign, so named after the Swedish Lotus F1 driver who succumbed to the disease at the end of 1978. As things turned out, logistics and sponsorship considerations meant the F1 race was replaced by the Procars. This had the distinction of being the last race of James Hunt's career.

The programme continued in 1980 with events supporting the Monaco, British, German, Austrian, Dutch and Italian grands prix, plus three separate events away from the F1 circus.

For Nelson Piquet, the 1980 series proved a lucrative and worthwhile consolation prize during a season in which he was narrowly beaten to the world championship by Alan Jones. By then, in the second year of a three-year contract with the Brabham team, Piquet was earning a basic retainer of only £25,000 for his F1 efforts, so his Procar earnings provided him with welcome pocket money.

Piquet took the same approach as Lauda, appreciating that handling the M1s required polish and precision rather than the hooligan attitude adopted by some rivals. The other outstanding performers that season were Jones and Stuck, who finished second and third in the series, plus Ligier's Didier Pironi.

Stuck earned the distinction of winning at Monaco, where Pironi's speed through the swimming pool section was absolutely awesome in practice, and these two men became embroiled in possibly the best race of the year at Hockenheim on the weekend of the German Grand Prix.

Willed on by vociferous support from his countrymen in the huge grandstands, Stuck became progressively bolder in his challenge for victory. But Pironi, who pre-dated Senna by a decade in terms of his 'ice man' image, neatly helped Stuck up one of the chicane escape roads at the height of their battle before sailing on to win. An enraged Stuck recovered to finish third, after which he had to be restrained as he unburdened his fury on the winner. Pironi, cool and impassive, simply stared at him in a mixture of bewilderment and contempt!

It was fun while it lasted, but at the end of the 1980 season the BMW Procar series came to an end. BMW now had more pressing priorities, having effectively taken Ken Tyrrell's advice and committed itself to the development of a turbocharged F1 engine.

More than a decade later, the JaguarSport Intercontinental Challenge means the XJR-15s pick up where the M1s left off. As Derek Warwick, Davy Jones and the other rich and famous tackle the streets of Monte Carlo this weekend, one can only speculate how long it will be before Tom Walkinshaw is back in the F1 pit lane with his own grand prix team. ■

The sound of Munich

There have been faster racing saloon cars than the BMW M3 – notably the Nissan Skyline GT-R and Ford Sierra RS500 Cosworth – but there has never been a more consistently successful and durable championship winner. From its debut season as a European and World Championship winner to the 1991 version which propelled Will Hoy, Roberto Ravaglia and Tony Longhurst to the British, Italian and Australian championships, the simple but efficient M3 has garnered over 30 titles worldwide.

The naturally aspirated, 16-valve, dohc, front-engined, rear-drive design has even won a World Rally Championship qualifying event in the age of 4x4 turbo cars. Similarly the M3 brought BMW unexpected rallying titles in Belgium, France, Spain and Yugoslavia, most of these due to the activities or component knowledge of Prodrive in Brackley.

The M3 has also been a commercial success, selling over 17,000 units (they needed only 5000 to qualify for Group A in 1986).

The opportunity to drive a representative example of the German Championship M3 racing breed (they won that title in 1987 and 1989) was not to be missed, especially when Briton Steve Soper would be on hand with fellow factory racer Altfrid Heger to explain and demonstrate the niceties of their 2.5-litre steeds. When a racing school proprietor (!) crashed the six-speed, ABS-equipped, 340 bhp racer it looked as though our reporter was headed for an early Italian bath. Yet BMW Motorsport racing co-ordinator, and former Grand Prix driver, Marc Surer had us back on track within 10 minutes in an older T-car, the amiable Heger's spare M3 being readied for afternoon action, its scuffed suedette steering wheel, lack of ABS and a dash full of aged dials betraying its *muletta* status.

Soper immediately proved that it could still lap within a second of the 51.996s lap record pace . . . with a passenger aboard! That record represents an average of just 70 mph, reflecting the presence of five second-gear corners and a first gear hairpin. The overgrown kart track at Magione, one-mile long and a couple of *auto-strada* hours up country from Rome, sunnily demonstrated the virtues of the M3 at speeds of up to 120 of the 180 available long circuit mph. The test car was fitted with one of 330 competition kits that BMW Motorsport has provided to over 40 countries. They will have paid the equivalent of £90,000 without a motor, or £140,000 complete with a BMW Motorsport power train.

The racing M3 is a car of enormous character, its four booming cylinders balanced to perfection by capacious brakes, simple suspension, and aerodynamic features. BMW Motorsport Racing Team Manager Karsten Engel admits: "We could make it better in some areas like the underfloor aerodynamics, but we are not allowed to do these things under German regulations, so the car will make its sixth racing season in 1992 largely unchanged." The big 3-series competition changes will come in 1993, when the company will field an M3 based on the current (E36) body shape, powered by a 2.5-litre, 24-valve version of the current 192 bhp 325i motor. Expected to give 250 bhp in road trim, it is now being readied for autumn 1992 production. A right-hand drive model will be sold in 1993.

For a preview, albeit without the extra aerodynamic appendages, watch the 1992 British Touring Car Championship, where Steve Soper will debut the new shape for his new British team, Vic Lee Motorsport. Prodrive is also scheduled to have one of the new bodies, complete with basic roll cage and body preparation from German fabrication maestro Matter.

We asked Steve to give us an insight into the differences between driving 2.5 M3s in Germany and 2.0-litre variants at home. He contrasted five key areas: weight, power, brakes, aerodynamics . . . and driving standards.

"At first," opines the now Monaco-domiciled Soper, "the British M3 actually feels quite slow. It's not just that it has 265 or so horsepower in place of 340 in Germany. You have to remember the weight. The less powerful UK car actually carries more weight. Of course it depends a bit how successful you are as a driver in Germany as to the actual weight you will carry, but the basic weights are 980 kg versus 1050 kg for the UK.

"Then there's the brakes. In Germany for 1991 we had ABS. In Britain, believe it or not, it was actually banned (it is now permitted – JW). You will also notice our German or Italian specification Evolution M3s have the front and rear adjustable spoilers. Not so in Britain. That makes an aerodynamic difference to the cornering ability I can call on, especially at the faster tracks."

During the 1990/91 winter a unique BMW Alfred Teves racing ABS braking system was conscientiously developed. Soper comments: "I tried the system back-to-back at Salzburgring and found *three* seconds in the wet, and it was better in the dry, too. It was a good thing BMW went for the system, because Mercedes had it for 1991 as well. I don't think their Bosch system is as good as ours in several important respects."

As to the opposition, Soper underlines the talent of men like ex-Grand Prix racers Jacques Laffite, Johnny Cecotto and Hans-Joachim Stuck in German fields full of rapid conductors. There were up to 15 drivers capable of winning in the German series, whereas the nascent UK equivalent naturally had a few less top runners recruited. Soper has kind words for the 'never-say-die' speed of John Cleland in the Vauxhall and Andy Rouse in his Toyota. However, since we spoke, the quality of the UK series has been augmented. Marque interest comes from BMW, Vauxhall, Toyota, Mazda, Peugeot and Nissan, at the very least . . .

Meanwhile, the Germans are looking at a leaner year than of late, with 30-car grids from only three manufacturers. The absence of Ford and Opel threatens to turn their year into another Audi rout with BMW and Mercedes

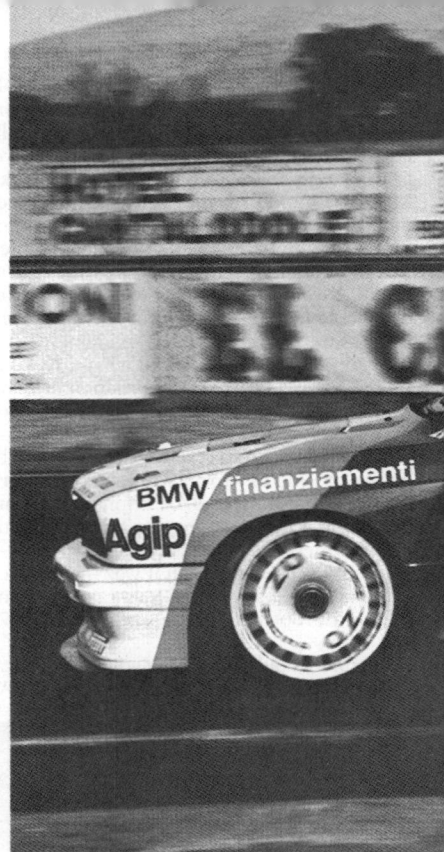

The M3 has collected over 30 saloon car titles . . .

squabbling over the Championship pieces.

A conventional Getrag five-speed gearbox has synchromesh operation, but a number of pure racing six-speed alternatives are used by BMW competitors, including products from Hollinger in Australia, Getrag and Prodrive. I have experience of the Prodrive unit, and liked it very much, but I am told the Hollinger is lighter and faster to deploy.

We were seated by Recaro and restrained by

BMW M3 GROUP A
(Prodrive BTCC specification)

ENGINE	
Location	longitudinally front-mounted
Cylinders	four in-line
Bore × stroke	93.4 × 72.6 mm
Capacity	1989 cc
Compression ratio	not available
Valve gear	dohc, four valves per cylinder
Power	275 bhp/8750 rpm
Torque	190 lb ft/7500 rpm
TRANSMISSION	
Type	six-speed manual, rear-wheel drive
BRAKES	
Front/Rear	ventilated discs all round
WHEELS/TYRES	
Techno 8J×17	Pirelli
DIMENSIONS	
Race weight	1050 kg
ESTIMATED PERFORMANCE	
0-62 mph	5.7 sec
Maximum speed	170 mph

well mannered. From the start I used 9000 of the usual 9500 rpm, but during the learning curve I took advantage of its good nature and broad torque spread to pull from 5000 rpm in the lower gears. Haul it back into first for the tightest corner and the natural rev band seems to lie between 6000-9000 rpm, but this first gear section is followed by a second-gear right onto the main straight. Even the stars back off here, so most regularly exploit the sub 5000 rpm torque.

The sound levels are modest by racing standards, but exciting nonetheless. It's fun to feel the large four smooth out above 5000 revs point, yet sheer speed is not overwhelming. You might touch 115-120 mph. Until the end of the main straight, that is, when fifth is held briefly before the demands of a tight right-hander. This was the most vulnerable point of a lap for a stranger. Even the works drivers occasionally dived at the apex, still changing gears furiously. At first the choppy ride and handling was composed, then the old M3's rear wheels start to kick out of line in a manner akin to the racing RS500 with which I was more familiar.

six-point Sabelts. The dashboard looks simple at first glance, a 10,000 rpm Stack tachometer placed adjacent to a transparent pane. "That rev counter tells them all that you have done and this small readout (behind the innocuous blank pane – JW) can tell of five or six things, including water and oil temperature, oil pressure, all things like this. We keep it on the water temperature for a race, because its needs to be kept in 85 to 90 degC for the best power," explains Heger.

BMW M3 GROUP A
(Schnitzer GTCC specification)

Walton at the controls. The onlooking Steve Soper is the only Briton currently earning a living in the GTCC.

ENGINE	
Location	longitudinally front-mounted
Cylinders	four in-line
Bore × stroke	95.5 × 87.0 mm
Capacity	2493 cc
Compression ratio	12 to 1
Valve gear	dohc, four valves per cylinder
Power	340 bhp/8750 rpm
Torque	214 lb ft/7500 rpm
TRANSMISSION	
Type	six-speed manual, rear-wheel drive
BRAKES	
Front/Rear	ventilated discs all round
WHEELS/TYRES	
BBS 9Jx17 (f)/9Jx18 (r)	Yokohama
DIMENSIONS	
Race weight	980 kg
ESTIMATED PERFORMANCE	
0-62 mph	4.8 sec
Maximum speed	180.1 mph

The catalytic converter is compulsory for this formula and it makes a big difference to drivers. On a hot summer day it can reach between 50 and 70 degC in the cockpit, because the converter is under the bare steel floor. Heger continued with an A-Z tour of the myriad warning lights. "The red one means you will have to stop, there is no oil pressure. The yellow ones are for the petrol tank, or the generator," he explains, although there are also lights to monitor ABS in the current frontline cars and a supplementary petrol pump. Within the white steel walls and Matter roll-cage protection lurks an old-fashioned racing car.

By the transmission tunnel, twin levers and cables lead fore and aft. "These are to operate the front and rear roll bar loads," smiles Heger. "Normally we would run with 70 per cent for the front bar, 30 per cent at the rear." Further cockpit sophistications include the usual brake bias hand wheel, an array of popout electrical fuses, a diagnostic plug for the ECU and the selector for that dashboard display of vital motor functions.

At some 136 bhp per litre, the M3 remains

As I started to use more opposite lock, the "in" board came out from the pit wall. Had I exceeded their hospitality? No, the tyres really are shot. Now I get a ferocious driving display from Soper, which produces a time despite visibly worsening grip. Ridged kerbs reverberate beneath the slewing BMW as Soper-the-Chauffeur urges more speed from flayed rubber. Rumble strips advance and retreat beneath hard worked Pirellis: tyre squeal emerges as the healthy motor and brakes are pressed to their limits. All give us an insight to what being a factory BMW driver really is all about.

Hard, but exhilarating, graft.

Soper sets a 54.77 sec best, two-up, where my best had been 57.56 sec on a solo run. The 2.79 sec gap is primarily accounted for by my plump cowardice, especially under braking at the end of the fastest straight. I am grateful to BMW Motorsport and Steve Soper for the time they allowed us to gain a unique insight into the professional saloon car driver's working environment for the '90s. I would still swap its centrally heated charms for this confounded wordprocessor . . . **JW**

We did our damndest to avoid France. Fearful of being held captive on some obscure autoroute by hideous garlic-chomping anarchists masquerading as lorry drivers, we'd booked a ferry crossing from Dover to Ostend.

This wasn't done lightly, for the Tomalin stomach turns queasy during the opening credits for the Poseidon Adventure, and the crossing to Ostend lasts *four* hours. But we had a long way to go, a lot to do, and precious little time. And on the day we made our ferry booking the whole of France was in the grip of these articulated thugs protesting about not being able to drive through red lights at 120mph without being brought to book. Or something.

Monday, departure day, and — *coup de théâtre!* — the French authorities (if that's not a contradiction in terms) had finally seen sense and sent in tanks to tow away the lorry drivers and push their wretched juggernauts into the fields. Which was just as well, because we missed our 10am ferry to Ostend by five minutes and were directed onto a boat bound for Calais.

Thus we were spared three hours of being tossed around in the Channel, sped through France at a rate of knots, and were past Ostend long before 'our' boat was due to dock.

Even better, the crossing was as smooth as could be, so the Tomalin stomach not only kept its contents to itself but added to them. We are talking cooked breakfast in the ship's restaurant.

We are also talking £8.20 for a thimble of orange juice, a cuppa, and a very average fry-up. And you thought piracy on the high seas was dead. To add insult, coffee arrived in a teapot. 'We're a bit short of coffee pots today,' said the steward, who was more concerned with herding all the smokers into one square yard of restaurant. A bit short of coffee pots *today*? Where did they go? (I'm sorry sir but we're a bit short of lifeboats today. And all you smokers, you're in that biscuit barrel over there. And you'll have to buy the biscuits first. So that'll be ninety-five pounds. Yo ho ho.)

Five countries in five hours?
Two people and a whole heap
of luggage? Sounds like a job
for Supersaloon!
Peter Tomalin drives the latest
BMW M5 across Europe

Photography: Derek Goard

Dial M

Dial M

But we were happy to have gained a couple of hours, and happier still when the M5 began to stretch its legs across the long miles of dreary northern French countryside that reminded me of the Cambridgeshire I'd left behind that morning.

This is the third incarnation of the M5, and the list price is now a cool £48,950. Visually, it's virtually indistinguishable from the previous car, but important changes lie beneath.

To give it more grunt, BMW Motorsport has bored and stroked the unique 24-valve straight-six (no relation to the six in the 535i) from 3.5 to 3.8 litres. With bigger valves and higher compression, and with the fuel injection and electronic ignition now under new Motronic management, it churns out more power (340bhp against 315 from the old car) and more torque (295lb ft versus 265).

To make it ride and handle better, in comes new 'adaptive' suspension, which means the damping rate adjusts automatically to take account of road speed, steering input, acceleration and body movements.

Clever stuff — and it works. The ride is remarkably good for such a sporting device, especially one wearing such wide, shallow rubber — 255/40 ZR17 Michelin MXX2s at the rear if you specify the optional

● **Handsome new alloys distinguish latest M5**

● **Right: the ultimate driving environment?**

128

'Nurburgring' pack (£1365) with 235/45s at the front. Compliant enough to cosset occupants on a long haul, firm enough to prevent any float or wallow, it is the best sort of compromise. Flick the 'Nurburgring' switch on the dash and you can set the dampers permanently to firm, but the ride becomes more joggly with no obvious benefit to the car's handling.

We stopped in Dunkirk, when Goard spotted the sort of decaying building that makes photographers go all wobbly

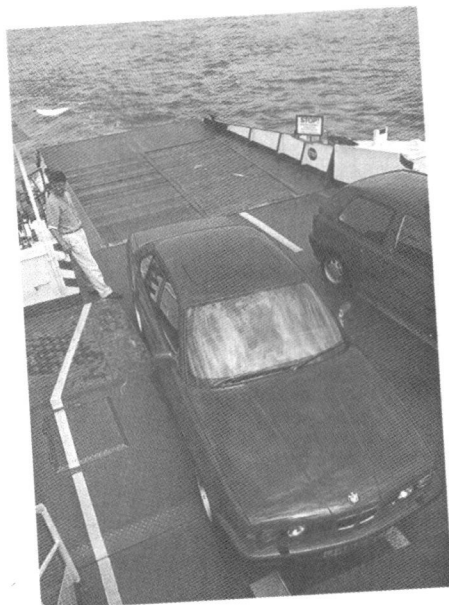

● Tomalin and M5 take a breather on Rhine ferry

and excited, and in Antwerp, where the vast docklands promised more photo opportunities, but otherwise France and Belgium and Holland slipped painlessly by — the Belgians realised long ago that few people wanted to stop there, and so built some of the best motorways in Europe.

We crossed the German border long before nightfall, marvelling at the ease of it all. Five countries in five hours (if you ignored the photocalls) and the M5 was still supping the same tankful of supergreen from the early-morning stop at Penny Farthing services on the M2. It lasted 362 miles at a rate of 23.19mpg, which is pretty good going for a hefty saloon that accelerates from 0-60mph in under six seconds, though the next tankful would see the figure dip below 20mpg.

Which brings us back to that engine. Now, BMW says that 75 per cent of peak twisting power (a more than useful 221lb ft of torque) is available at just 1800rpm, and the big six (the biggest six in the company's history, in fact) certainly feels tractable. It'll pootle along in fifth at 40mph and pull away without hesitation. But I reckon you have to work it much harder than that to extract anything like its true potential.

Slice the slick five-speeder down a couple of gears, and as the revs climb past 4500 it's as though it gets a new lease of life, singing to the 7000rpm red line like a Motorsport engine should. It sounds very busy, very mechanical — the slightly menacing, uneven throb of the exhaust at tickover becomes a distant growl as you pick up speed — and you sense it could take this treatment for just

as long as you wanted to dish it out.

But although the performance on tap is formidable, the M5 is a pussycat to drive. Most BMWs have a very light, precise, user-friendly feel to them. The M5 retains the precision and the friendliness, but all the controls feel beefed up, from the Servotronic steering to the weightier clutch and gearshift. Even the throttle has a stiffer action, though this is not good news — in fact it's one of the M5's few weak points, being difficult to apply in small measures.

We'd travelled to north-west Germany, where industrial towns like Duisberg, Mulheim and Essen merge into one great, grey sprawl, to spend the next day and a half sampling the latest products of Mercedes tuner Brabus (see separate story, page 100). Now, when you arrive at a Mercedes performance specialist in the new BMW M5, it doesn't go unnoticed. In fact Brabus's sales executive insisted I took him for a drive, and shortly we arrived on one of the all-too-rare stretches of unrestricted, lightly-trafficked

129

Below: Motorsport straight six has grown to 3.8 litres. There's plenty of low-down torque, but it loves to rev and rev...

BMW M Power

autobahn left in Germany.

The M5 didn't disappoint. In no time we were flying at 140mph, and when the two-lane stretched out into the distance we went all the way to 155mph, which is about as fast as BMW lets you go. And this is the truth: the M5 feels as relaxed at 155mph as most cars do at 75. Engine noise and road roar are subdued, the rush of air round the A pillars and door mirrors being to the fore, and the car is rock-steady. The man from Brabus was rather impressed.

Tuesday evening, and it was time for more photographs. As Goard fired away on the banks of the Rhine, I drunk in the M5's handsome lines. The hunks of rubber give it a

more aggressive stance than the average 5-series, and the new alloy wheels, the only real distinguishing feature of the new M5, have spokes like sculpted blades. But this is no muscle-bound Rambo, more 007 in a well-cut suit. I couldn't help thinking this is the sort of car Jaguar should be making in the 1990s. Perhaps when the Mk 2 is reinvented . . .

Wednesday afternoon, and after saying our farewells to the guys at Brabus, cornering shots were next on the agenda. So we headed north, away from the urban-industrial sprawl. And just as I was beginning to think the M5's place was in the fast lane of the autobahn, it threw down another ace. This chassis has all the fluidity, all the balance you could ever

crave in a saloon. Even when you're not in the mood for heroics it will cover country roads at a storming pace, calling on considerable grip and quick, precise steering.

And when you want to whoop it up, the M5's game. Maybe it doesn't communicate through the steering as the best sports cars should, but cornering attitude can still be dictated by the throttle and, if you overcook it, breakaway at either end is wonderfully progressive. The brakes, huge ventilated discs all round, with ABS, are simply magnificent.

Time to go home, and we had a race on our hands to catch the quarter-to-midnight ferry back from Calais. We needn't have worried. The M5 saved its best until last.

A diversion near Dunkirk saw us on single-carriageway roads, thick with juggernauts (all of them moving, praise be, but not at the pace we needed to keep). With Goard acting as look-out, estimating the closing time of oncoming headlights, we powered past two, three, even four of these leviathans in single bites, holding third gear and hearing the revs soar to the heavens. There were sweeping corners too, taken at unreasonably high speed, and when the road surface became uneven, the suspension soaked up everything that came its way. It was a virtuoso performance.

Other supersaloons offer more power, more grip, more road presence, more luxury, but none of them has such a well-rounded character, not one of them adapts so readily to the driving style of your choice. The M5 really is your flexible friend.

It was 2.30am when I finally pointed the M5 back up the M2 for home, and the July sky was lightening at the corners as I tumbled out in slumbering Cambridge. All those countries, all those miles, but somehow the M5 made sense of it all. ○

Roadwords

BMW Motorsport only builds 12 of these cars a day. And that's on a good day

Best sporting saloon in the World? Make a claim like that and you are open to attack. You can completely dismiss the 'well, at 50 grand it ought to be' reaction because those people are obviously not in the market to buy, and so can be ignored. The serious criticism will come from those who know, because they own challengers to that title; and from impecunious journalists. Like **CCC** Editors. Like me.

At the end of a bad week, being handed the keys to an M5 to drive home does help to soften the blow. Friday nights might see you 'celebrating' the coming weekend with a swift pint, or sitting inside a BMW taking in the intoxicating smell of quality leather. Both experiences act like drugs and both have their value – although not at the same time. In the case of this BMW M5 the leather interior is bright, cream coloured and squeaky clean. Why cream? Do people who can afford these cars never get mud on their shoes? Do their children never eat chocolate?

It's all a strange combination. The excellent paint job on the bodywork is called Daytona Violet (?) apparently. Robinson's Ribena is much closer, but it clashes quite abruptly with the Cadbury's caramel cream interior. Who chose this combination we wondered. After even the shortest drive we understood. The visuals of this particular model have been done to try and shock you away from the vehicle's dynamics. It's simply daring you not to like the thing.

This latest M5 is actually the third generation model. This means that subtle changes to the six cylinder engine have increased the power from 315 to 340bhp, with torque up from 265 to 295lb ft. The characteristics of the engine means that 75 per cent of this torque (221lb ft) is available from only 1800rpm. Slot this car into fifth gear and you can do 20mph.

This characteristic, coupled with a lighter clutch, does allow for smooth driving in even the heaviest traffic. What it doesn't – thankfully – do, is divorce you from the impression that you have one seriously powerful engine under the bonnet. The M Sport engineering department of BMW is not yet staffed by anaesthetists, or Japanese engineers. Yes, of course it's smooth, svelte and all the other clever words you can read by the Sunday supplement motoring correspondents. But it also lets you know it's there.

The BMW feels so solid, you wonder what other manufacturers do when they build cars. It has such a low key appearance – monster wheels and tyres, subtle additional body kit apart – that this is just another BMW 5 Series and let's face it, recession or no, there's a lot of them about. Even the wonderful M Sport styled steering wheel has gone to be replaced by the air-bag variety. (I still cannot drive a car without continually wondering if it is going to explode in my face.) If it doesn't sound too crass, you *feel* the presence of an M5. If you think **CCC**'s wandering dangerously close to that Sunday supplement garbage again, apologies. Driving this car is quite special.

One of my runs took in some open southern England B roads; dips and dales, cambers and pot-holes. And Sunday drivers. The M5 was magnificent. Cool, calm and collected behind these assorted dream wheelers, you could unleash the thing at an overtaking manoeuvre and surge pass. At one point on my journey, this involved a quite noticeable climb. The rapidly approaching top of a blind brow meant that discretion was necessitated by our velocity; lift off, brake. Classic error of judgement since this was done at the peak point of the brow while the car was rising and therefore going light on the suspension. It was not a problem. I was still in control of the car and the car was still in contact with the road. We slowed (the M5 has awesome brakes), checked the route, confirmed it to be safe and clear and accelerated again. Marvellous stuff and a great compliment to the car's electronically controlled adaptive suspension system.

This is actually BMW's third variant to the Electronic Damper Control system (EDC III). It responds to steering input, acceleration and braking as well as body movement and alters the damper settings front and rear to suit. Nigel Mansell would be proud of the car although that is a little OTT since the system is adaptive, not truly active. It does make you wonder, however, and go into journalist mode. If Ron Dennis loses Honda engines for next year's Grand Prix season and turns to BMW (since they are building the engines for the McLaren F1 road car) will the German company also bring along electronic suspension knowledge? Hmmm. At rather slower speeds than an F1 car, bouncing over London's sleeping policemen, EDC III gets a little confused but on the move, at speed, it is impressive and gives the driver great confidence. The Servotronic speed-sensitive steering was particularly impressive.

In pure performance figure terms the M5 will manage the 0-60mph sprint in under six seconds and gallop along to the 155mph mark before the Motronic engine management system calls a halt. Impressive though they are, the figures only scratch the surface of how the car really *performs*.

BMW Motorsport only builds 12 of these cars a day. And that's on a good day; it can be less. In the UK, they are only likely to import 60 models a year. If you bought one, it would be difficult to change it. Ask yourself what more you could want? The build quality is there; the BMW M5 may depreciate but it won't deteriorate. The glorious 17 inch wheel and tyre combination have enough pose value alone to scowl at most two seater supercars. And this one will take five adults and/or loads of luggage. It's more refined and got a great CD player. And that performance.

If it's not the best sporting saloon in the World, it's bloody close. ∎

NIGEL FRYATT

● 3.8-litre engine packs a mean punch with 340bhp on tap giving a top speed of 155mph and 0-60 in under 6secs

BMW M3

THE "M" MEANS MORE: MORE POWER, MORE PERFORMANCE, MORE FUN

by Bernard Cahier

There's one sure bet in the automotive industry: No matter what new vehicle is introduced, there'll be at least a half dozen after-market manufacturers waiting with products to make it perform better and look snappier. Sometimes it works; most times it doesn't. Unless, of course, the metamorphosis is done by the original manufacturer. Perhaps one with a long history of racing success and a proven record of producing exceptional performance products. We could only be talking about BMW, and the good folks at BMW Motorsport GmbH once again have transformed the already marvelous 3-Series coupe into the M3, a Porsche-eating scene-stealer.

Putting tights and a cape on its models is nothing new to BMW. It all started with the M1 in the late '70s and is ongoing with the M5 and soon-to-be introduced in Europe 850CSi, which incorporates some M5 technology. The '93 M3 is the second iteration of this line, with the first M3 debuting in America in 1987.

Armed with a 24-valve 3.0-liter DOHC inline six, the new M3 boasts a horsepower rating of 286 at 7000 rpm and 320 pound-feet of torque at 3600 rpm. In contrast, the standard 325is offers 189 horsepower at 5900 rpm and 181 pound-feet at 4200 rpm, while the old four-cylinder 16-valve M3 kicked out 192 ponies at 6750 rpm and 170 pound-feet of torque at 4750 rpm. With nearly 50 percent more power and 88 percent more torque than its predecessor, the M3 possesses some mighty impressive acceleration. Our '93 Import Car of the Year nominee 325is ran 0-60 mph in 7.4 seconds; BMW's claim for the M3 is 6 flat. BMW also asserts the M3 has the highest specific output per liter in the world for a naturally aspirated street-legal production vehicle: 95.7. The Acura Integra GS-R comes next, at 95.4. As for top speed, there's an artificial limiter set at 155 mph, but it definitely feels as if the M3 could climb into the 160 range.

A tweak here and there under the hood is responsible for this chest-pounding performance. The use of BMW's Variable Camshaft Control explains the increased torque throughout the midrange; an increased bore and stroke, new pistons and valves (still four per cylinder), and a compression ratio bumped up from 10.5 to 10.8:1 put the *mmmmm* in the M3. In addition, a beefier suspension with firmer dampers and anti-roll bars, reinforced springs, and modified axle geometry at all four corners produce the expected results.

On our drive around the breathtaking island of Mallorca in Spain, we experienced all the pleasures of BMW Motorsport's handicraft. Launching from a stop produces instant response; the precision-made screamer starts quickly and easily fills its lungs all the way to redline, producing power in a continuous, breakneck ascent. The gear ratios are well spaced, smooth in operation, and excellently matched to the engine's torque.

Likewise, handling was comparable to the best from Japan and those "other" European manufacturers of sports cars. On the tight curves of narrow mountain roads, the M3 was constant and predictable, with little body lean.

The rack-and-pinion steering feels precise, and the anti-squat/anti-dive geometry made a noticeable difference in stability on the S-curves. You can induce oversteer if you go into the corners a bit hot, but the 50/50-percent weight distribution and a self-locking rear differential can help bring you back under control with minimal hair loss.

To get you back down to earth, the M3 has increased the rotor diameter of the four-wheel vented disc brakes by over 1 inch both front and rear, as well as increased the front/rear thickness. Toss in BMW's superb ABS system, and you have a car with absolutely magnificent stopping power under any conditions.

You'd think this increase in performance alone would make the Baby "Bam-Bam" Bimmer good enough to show to the public, but you've got to dress it up before you can take it out. BMW Motorsport clothed this youngster in one of its three exclusive M3 colors: Dakar yellow. If it's too bright, try Mugello red or Avus blue metallic. Color-matched side skirts, a front spoiler with a mesh air intake, modified rear apron, specially styled sideview mirrors, plus BMW Motorsports original 17x7.5-inch cast alloy wheels wearing 235/40ZR17 tires make up an exterior package that screams speed. Top it off with the required M3 trademark tri-color badge, and few who see it will challenge you on the road.

Inside, the M3 gets new, extra-support sport bucket seats adorned with a colored BMW Motorsport raindrop-pattern cloth mixed with Amaretta, a simulated-suede seat cover. For the instruments, BMW improved the speedo with a 175-mph top-speed mark, added an oil-temperature gauge, and in the tradition of BMW racing, changed the color of all the indicator needles to red.

What's this super-hero upgrade going to cost? No firm prices have been set, but BMW estimates somewhere around $60,000, roughly twice the price of the factory 325is. BMW still hasn't confirmed whether the M3 will be coming to America, but if it does, you can bet it will be in limited numbers.

With the new M3, BMW once again has proven it remembers well its heritage, understands what the enthusiast wants, and has a knack for turning visions into dazzling reality. **MT**

The M3 gets upgraded sport bucket seats and red indicator needles on the instruments. Special alloy wheels and body-color side cladding create a sporty-looking package.

TECH DATA

BMW M3

GENERAL/POWERTRAIN

Body style	2-door, 5-passenger
Vehicle configuration	Front engine, rear drive
Engine configuration	Inline 6, DOHC, 4 valves/cylinder
Engine displacement, ci/cc	183/2990
Horsepower, hp @ rpm, SAE net	286 @ 7000
Torque, lb-ft @ rpm, SAE net	320 @ 3600
Transmission	5-speed manual
Axle ratio	3.15:1

DIMENSIONS

Wheelbase, in./mm	106.3/2700
Length, in./mm	174.5/4433
Base curb weight, lb	3219
Fuel capacity, gal	17.2

CHASSIS

Suspension, f/r	Independent/independent
Steering	Rack and pinion, power assist
Brakes, f/r	Vented discs, vented discs, ABS
Wheels	17 x 7.5, alloy
Tires	235/40ZR17

PRICE

Price	$60,000 (est.)

Worth the weight?

Behind the reborn BMW performance flagship lie at least four years' development graft and 50 prototypes. Replacing the familiar old four-cylinder warrior is a very much smoother six-cylinder M3, but the 155 mph newcomer is a very much heavier, plusher machine. It is more akin to a replacement for BMW's old six-cylinder CSi coupé. This time, British customers *will* be offered rhd from May 1993 at predicted prices in the £33-34,000 band, depending on the exchange rate and what items from the considerable options list are incorporated as part of standard UK specification.

No agreement has yet been signed to import the M3 to North America, where predicted prices look uncomfortable in a Japanese-dominated market.

Nothing but the badge and its origins in BMW Motorsport really link the new M3 to its predecessor. The 0.32 Cd body is based on that of the current 3-series coupé and much of the running gear has been developed from the 24-valve, dohc 325i.

Furthest from its 325i base is the engine. A new iron block and two-part alloy cylinder head are, statistically, very different. The 24v dohc principles are substantially modified by a variable valve timing system (VACC for English speakers, VANOS is the acronym used by Germans) that works on the inlet camshaft only. Its electro-hydraulic actuation modifies camshaft action between 80 and 120 crankshaft degrees and allows exceptional torque to accompany the generation of 286 bhp at 7000 rpm.

BMW credits its latest M50-coded three-litre (it's actually 2990 cc, 86 × 85.8 mm) with a world record in bhp per litre (95.6) at this capacity, though Honda might like to query that. The Civic VTEC manages 98.8 . . . BMW has, however, overcome Porsche's previous benchmark for torque in the normally-aspirated, three-litre class. The M3 pumps out 236 lb/ft at 3600 rpm. The most remarkable facet to that torque capability is that the peak figure is sustained all the way to 5800 rpm. Furthermore, the engine remains remarkably suave at either end of its 7280 rpm range, be it 700 or 7000.

The unit is mated to a 525i ZF five-speed, a sextet of ratios being ruled out by Germany's target price ceiling. The bulky motor is installed with at least two of its in-line cylinders forward of the axle line, and one can see why BMW and the organisers of the German Touring Car Championship came into such bitter dispute over the engine location. The company claims that 50-50 weight distribution is retained on the road, but it does not look like that in the flesh; before withdrawing from further GTCC involvement, BMW Motorsport wanted to run a lowered and relocated (moved back by 15 cm) version of the motor.

For road customers the 325i chassis has been substantially overhauled, but the principles of MacPherson strut front and Z1-derived multiple-link back axle remain, albeit with typical detail reinforcements. These centre on a 1.22 in reduction in ride height after the adoption of replacement gas-filled damping, shorter and stiffer coil springs, thicker front anti-roll bar, stiffened suspension mounting joints and a ZF limited slip differential operating a 25 per cent torque preloading. The steering features a variable ratio that is not offered elsewhere in the 3-series range. The back trailing arms are physically strengthened and 850i Coupé wheel bearings are adopted, all to resist the ravages of cornering forces up that have been measured at up to 1.2G.

The biggest handling advantages come from more obvious modifications: 17 in wheel diameters are adopted with 7.5 in rims (8.5 in is a forged alloy option) and 235/40 ZR covers from the Michelin MXX3 range are prominent. Behind such large diameter wheels lurk equally impressive vented discs, effectively monitored by Teves ABS.

BMW makes much of the new M3's stopping abilities. It claims that it takes just 8.8s to go from rest to 62 mph and back again; six of those seconds are spent gathering velocity, and just 2.8 scrubbing it off.

The body has none of the old car's extended wheelarch character; even a rear spoiler is only allowed as an option. The most notable M3 panel is an extended front spoiler with under-bumper grille, but detail modifications also abound, such as prominent side sills and an extended, stylised rear apron, which wraps around stubby twin exhausts. BMW Motorsport sales personnel did feel that more M3 identity was needed, and the 'racing' mirrors, badges on each door and the kickplates reflect that lack of confidence. Originally they were going to put an M3 badge within the sacred kidney grille, but corporate defenders of the faith vetoed that intrusion, leaving the bootlid as the only place to put such a motif.

The interior is not inspired, but it does carry an individual Motorsport trim alongside the ponderous air bag steering wheel (intricately hand-stitched in bright threads around the inside circumference of the rim) and an oil temperature gauge continues to replace the econometer of the usual 3-series. Red needles record 7300 as the limit on the 8000 rpm scale and 270 km/h (168 mph) as the maximum indicated speed, but, as with all current high-performance BMWs, an electronic device interferes at 155 mph.

The engine has all the hallmarks of BMW breeding, whirring along at low rpm, where there is plentiful torque available, and gradually taking on the note that encourages lazy colleagues to recall the "turbine smoothness" of a BMW straight six.

Whether humming along melodically at 87 mph (less than 4000 rpm in top), or sprinting between second gear curves at 7000 rpm, the seductive engine is reason enough to buy. BMW engineering legend Paul Rosche told journalists: "I know of no better engine." We would endorse that for the combination of civilisation and accessible power.

Other M3 abilities that set new benchmarks are those enormous brakes and a beautifully absorbent ride, especially when you consider the 40 per cent aspect ratio tyres.

What we were not so impressed by was the communication between car and driver over roads with differing degrees of adhesion. On dry surfaces there is enormous grip, and the Michelins awaken the steering from its apparently well damped torpor to tell you that initial understeer has been traded for old fashioned power oversteer. When, as for our trip, surfaces go from dry to damp to wet and back to dry, the M3 is far from convincing. The driver – despite adjustable shoulder and headrest sections to the unique seats – is not made to feel at one with the car. The steering fails to relay either the drop in bite being achieved by the front wheels, or the disquieting speed with which the rears will now want to overtake their forward comrades. It was not the kind of hair-raising experience, or general unease, that 911s and Skodas used to generate in their original formats, but the new M3 is hard to balance in its new, heavyweight guise.

I am not advocating a traction control device, which Motorsport eschewed but is now working on again for this model, but the kind of alert feedback that characterised its 527 lb lighter four-cylinder progenitor. The new M3 can't match the old when it comes to seeking pleasure on a twisty road. It was a similar story when the 3.8-litre M5 supplanted the original.

The truth may be that journalists and purists enjoy these original Motorsport devices, but the buying public always want further equipment. Such bulk soon ruins handling pleasure, as well as putting a dent in acceleration curves and making affordable fuel consumption a challenge. In the latter respect, BMW points to a likely overall average of 31 mpg; we recorded a best of 18.7 and the most probable UK averages will be close to the urban quote 21.7. It runs on 98 RON super unleaded.

I would put the new M3 on my personal wanted list, but I wonder how many others amongst the 17,000-plus buyers of the first M3 will be amongst the 24,000 anticipated customers (over the next five or six years) for the new one? I suspect many more old 6-series clients are going to be interested, in which case the machine could have been most accurately described as a 3.0 CSi.

That would prevent purchasers gaining the false impression that the new M3 is as raunchy a driving machine as the original. **J W**

M3

BMW's best sells its soul

This new M3 is probably the fastest road car BMW has built. So why is Andrew Frankel unconvinced?

First, the good news. When the new BMW M3 goes on sale next May, it will have right-hand drive and, at £33,000, will cost some £1500 less than BMW asked for the last M3, the Sport Evolution of 1990. It is also very, very fast. Its creator, BMW Motorsport managing director Karl-Heinz Kalbfell, says it could be geared to do 175mph with no more power from its new three-litre six-pot engine.

The bad news? That can wait for now. If you want a clue, cast an eye over its shape. Doesn't *look* much like the old M3, does it?

Then again, maybe you wouldn't expect it to. The self-conscious '90s call for purity of line to replace the bulging bewinged machismo that so distinguished the M3 in the carefree '80s. The lines of this M3 are as unadulter-ated as you could wish. In a photograph, you might mistake it for a 325i on over-ambitious wheels.

But not in the flesh. For a start, it is 30mm lower. And BMW has been careful to exploit this to the full, concentrating all changes to the bodywork at ground level. Thus, you will find a deep front spoiler, new side skirts and rear apron which contrive visually to lower the car further still. Otherwise, save a scatter of M3 badges on the rump, sides and kickplates and special aerodynamic wing mirrors, you have to look inside before you'll find further resurrection of the old M3's memory.

New front seats with headrests that move in unison with the radical adjustable shoulder supports mark this out to be no ordinary 3-series, as does the suede-like upholstery. The familiar M-badge looms large on the instrument panel and gear lever as before and, in time-honoured M-series style, the instrument needles are red and an oil temperature gauge replaces the economy scale. An interior that honours the memory of the M3 then? In the main, yes. After all, you can't quibble about swapping the Motorsport steering wheel for one with an airbag, can you? Not when BMW will put it back if you ask

Dash is little different — why alter a winning formula?

It says M3, but the driving experience is more 330 CSi

nicely. But take another look at that gear lever. M3s used to put first, not fifth, out on its own, in classic racing style. No more. It's a small point, especially as BMW says a conventional layout is the more popular, but it is significant.

Significant, too, is the sight that greets your eyes under the bonnet. Look for a gaily painted engine or scarlet plug leads and you'll search in vain. All you'll see are the words 'BMW M-POWER' standing proud as ever of the crackle black cam covers.

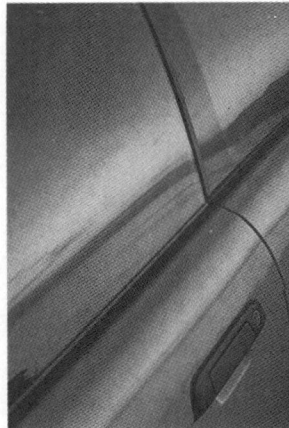

Sun sets for the M3?

What you have found, in fact, is an engine that vies with that of the Honda NSX for the title of finest six-pot powerplant made today. Developed from the twin-cam 24-valve M50 engine that powers 3-series and 5-series BMWs, it gains a new capacity of 2990cc and retains the VACC variable valve timing system (for VAriable Camshaft Control). The result is 286bhp at 7000rpm and 236lb ft of torque at 3600rpm. M-series aficionados will notice that the power is identical to that

Steering lacks the precision and feedback of the old M3. With up to 1.0g cornering on offer, it should be better

of the much loved 3.5-litre M635 CSi. As we shall see, this is no coincidence. BMW claims that its specific output of 96bhp per litre is a record for a normally aspirated car, (conveniently forgetting the Honda Civic VTi's 99bhp per litre) and calls the torque curve, flat all the way from 3600 to 6000rpm, Ayers Rock. It also points out that, at idle, it has more torque than the old M3's engine could muster at any speed.

The job of keeping such energy pointing in the right direction is entrusted to a chassis that, while sounding like the familiar 3-series cocktail of struts at the front and multi-link Z-axle at the rear, has been modified beyond useful comparison. Naturally spring and damper rates have been radically revised, but that is only the start. Modified track arms up front bring new geometry, the stub axles themselves are reinforced and the anti-roll bar diameter has grown. The rear benefits from similar revisions to its anti-roll bar and control arms, while the wheel bearings are borrowed from the 850i coupe to cope with lateral acceleration claimed to be over 1.0g.

Other changes include fitting a variable ratio steering rack and massive ventilated discs — they are over an inch larger than before and come with a larger master cylinder and rerated ABS. Power is fed to the road via a limited-slip differential and 235/40 ZR17 Michelin MXX3 tyres, on 7.5x17ins rims. If you want, 8.5ins rims can be specified at the rear. Never let it be said that BMW cuts corners.

Yet, despite such comprehensive behind-the-scenes activity, anyone who has driven the latest 3-series will feel at home as immediately as any old M3 driver will feel alienated. BMW showed its car to the world's press in Majorca, and as I motored gently from Palma to the mountains I struggled to see the difference between this and the 325i coupe that had been my transport for the previous week. Despite obviously shorter gearing, the M3 is quiet and refined, comfortable and sophisticated. ▶

139

◄ Just like the 325i. The controls require no extra input, the gearchange has the same swift action, the clutch still bites annoying high in its travel. True, the seats hold your body tighter and the engine, even when warming through, has a more urgent, higher note. But where was the thrill, that raw seam of inspiration that, in M3s of yore captured your heart and mind as soon as you saw the car and didn't let go until you were safely home again?

As soon as the oil temperature settled, I determined to find out. The first thing I discovered was that, point to point, this is probably the fastest car BMW has built. It would crucify an old M3, string up a new M5 and I'll spare the 850i's blushes. It scarcely matters what gear you're in: press the pedal to the floor and suddenly the horizon doesn't seem so far away after all. The engine, like that of the Ferrari 512TR, has a fascinating complex of whirrs and growls at low speed which, one by one as the revs rise, defer to make space for a single, diamond-hard howl from 6000rpm to the cut-out at 7300rpm. It may have six cylinders, but the heart and soul of the M3 lives and breathes in this engine as much as ever before.

The same cannot be said for the chassis and steering. The old M3's precision, its essence of all that is good in a racing car's chassis distilled into a civilised road machine, has gone. Turn-in, balance, feel and agility have been sacrificed at the altar of the great God of Grip. I believe BMW when it says the M3, given a decent surface (an infuriatingly rare commodity in Majorca) will generate over 1.0g in corners. I believe that this ability, the engine and the tireless brakes will make it faster point to point than all bar a rarified strata of cramped quasi-supercars. And I know all this can be achieved with four adults on board without upsetting their comfort.

But I also know that this is where the M3 has missed the point. It is not the first car to mistake pace for pleasure and it will surely not be the last. That is no excuse. It is, of course, still an enjoyable car in which to travel; the

Airbag is standard, sports wheel available on request

Dash is familiar save red needles and oil temp gauge

Exclusive M3 wing mirrors look wonderful

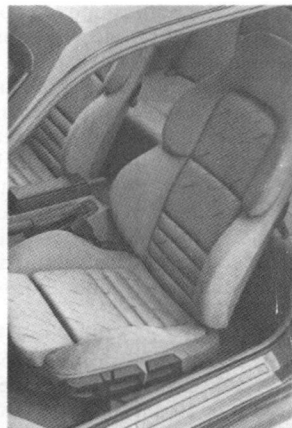
New seats for M3 only

difference is that now it matters less which seat you are in. You can appreciate the towering acceleration and cheek-rippling grip from the other side of the car just as well. And you won't have to contend with steering that is little swifter and no more communicative than that of a 318i and a chassis that, when it does let go, does so in more of a skid than a gently progressive slide.

The truth is — and BMW admits it — this car is no successor to the M3 at all. It would like you to think of it as a car cut from the mould of the M635 CSi (hence the 286bhp power output), a swift and efficient express, all wrapped up in a beautiful coupe shell. All these things it undoubtedly is. Anyone looking for a spacious, sophisticated alternative to a Porsche 968 or Mazda RX-7 will doubtless be delighted. I know I would be. But if you expect it to encapsulate the spirit of the true road racer as its forerunner did so well, you are likely to be disappointed. I know I was. And that's the bad news. ■